Green Energy and Technology

For further volumes:
http://www.springer.com/series/8059

Stephen O. Dean

Search for the Ultimate Energy Source

A History of the U.S. Fusion Energy Program

To Don Steiner,
In remembrance of our MIT days
and all the great work you
did for fusion over the
years!
Steve Dean
November 2018

Springer

Stephen O. Dean
Fusion Power Associates
Gaithersburg, MD, USA
fusionpwrassoc@aol.com

ISSN 1865-3529 ISSN 1865-3537 (electronic)
ISBN 978-1-4614-6036-7 ISBN 978-1-4614-6037-4 (eBook)
DOI 10.1007/978-1-4614-6037-4
Springer New York Heidelberg Dordrecht London

Library of Congress Control Number: 2012952107

Printed on acid-free paper

Springer is part of Springer Science+Business Media (www.springer.com)

I dedicate this book to my children, Stephen, Richard and Marcy (Rossi); to my grandchildren, Graham, Nicolas (Rossi), Spencer, Laura (Rossi), Claire and Eliza; and to my descendants, yet unborn.

Prologue

A journey of a thousand miles begins with a single step
–Chinese Proverb

In the summer of 1958, after completing my sophomore year as a physics major at Boston College, I received in the mail my monthly book selection from the Library of Science. It was Project Sherwood—The U. S. Program in Controlled Fusion, by Amasa S. Bishop [1]. It described a previously highly classified research program that had begun in the USA in 1951 under the auspices of the US Atomic Energy Commission (AEC). Bishop had managed the program for the commission during much of the 1950s.

In the Introduction, Bishop noted that "the practical resources of conventional fuels are dwindling at an alarming rate" and says it is "highly significant, and indeed fortunate, that a vast new source of energy looms on the horizon—the energy of nuclear fusion." He goes on to state, "If present efforts are successful, man will have found the ultimate solution to one of his most pressing problems. He will have developed a new and practical source of energy which will meet his needs, not for just the next hundreds or thousands of years, but for as far into the future as he can see."

Fusion is by far the dominant energy source in the universe. It is the source of light and heat in our Sun and the billions of stars that populate the heavens. On planet Earth, its energy has been tapped directly in the form of the hydrogen bomb, by far the most powerful explosive known to man. As the source of light and heat from the Sun, it is also the source of energy that permits life to exist on Earth. It is only natural, therefore, that scientists should seek ways to produce fusion energy on Earth in a form that could provide a long-term solution to increasing energy demand.

The reasons for pursuing fusion are as true today as they were in the beginning: it would provide a universally available, essentially inexhaustible, efficient fuel resource; it would provide an option to meet growing global energy demand; it would provide an energy source with attractive safety and environmental characteristics; and it would spawn new technologies for a variety of applications.

Fusion has sometimes been called "The Holy Grail" of energy, and its researchers likened to knights on a quest. Perhaps because of this, or perhaps it was just coincidence, the secret fusion research project in the USA during the 1950s was given the label "Project Sherwood." One of the early fusion scientists, Jim Tuck of Los Alamos, was often referred to as "Friar Tuck of Sherwood Forest." He was one of many colorful pioneers of the early fusion effort.

Critics of fusion, of which there have been many, laugh that the only thing sure about fusion is that it is always 20, 30, or 50 years away. Advocates, who fortunately have always outnumbered critics, often call fusion the "ultimate energy source."

After reading Bishop's book, I was hooked. For a young man trying to decide on a career path among the many areas of physics and technology, the choice seemed clear to me. This was a new area of science and technology with great potential to produce important societal benefits during one's working career. I made that choice and have spent the last 50-plus years pursuing that dream. After receiving a B.S. in Physics from Boston College in 1960, I went to the Massachusetts Institute of Technology, receiving an S.M. degree in Nuclear Engineering in 1962 under an AEC Fellowship. I then landed a job with the small fusion program management team at AEC, where I spent the next 6 years while simultaneously pursuing a Ph.D. in Physics at the University of Maryland. Those years are described in Chap. 3 of this book.

It turned out that the United Kingdom (UK) and the Soviet Union (USSR) had also independently begun research to harness this energy process of the Sun. A success of sorts came quickly in the form of the hydrogen bomb, but producing controlled thermonuclear reactions, or nuclear fusion as it is now more commonly called, remained elusive. The high-temperature, ionized, hydrogen gas (called "plasma") turned out to be much more difficult to contain in various "magnetic bottles" than scientists originally hoped. Consequently, at the Second United Nations Geneva Conference on the Peaceful Uses of Atomic Energy in 1958, the USA, the UK, and the USSR declassified their research. Bishop's book was commissioned by the AEC to coincide with this conference.

In this book, I provide a fusion concept primer, a personal perspective on the history of fusion research as I participated in it (others may have a different perspective), and look at the energy "problem" and how and when fusion might contribute to its solution. I address the questions: what is fusion, why has it always seemed just out of reach, will it succeed, and can it rightly be called "the ultimate energy source"?

Gaithersburg, MD, USA

Stephen O. Dean

Acronyms

AEC	US Atomic Energy Commission
ANS	American Nuclear Society
CTR	Controlled Thermonuclear Research
DCTR	Division of Controlled Thermonuclear Research
DOE	US Department of Energy
EPRI	Electric Power Research Institute
ERAB	DOE Energy Research Advisory Board
ERDA	US Energy Research and Development Administration
EU	European Union
FEAC	US Fusion Energy Advisory Committee
FESAC	US Fusion Energy Sciences Advisory Committee
FPA	Fusion Power Associates
FPAC	DOE Fusion Policy Advisory Committee
IAEA	International Atomic Energy Agency
ICF	Inertial confinement fusion (aimed primarily at weapons physics)
IFE	Inertial fusion energy (aimed at energy applications)
IFRC	International Fusion Research Council
JET	Joint European Torus
LANL	Los Alamos National Laboratory
LBNL	Lawrence Berkeley National Laboratory
LLNL	Lawrence Livermore National Laboratory
LMF	Laboratory Microfusion Facility
ITER	International Thermonuclear Experimental Reactor
MFAC	US Magnetic Fusion Advisory Committee
NIF	National Ignition Facility
NNSA	National Nuclear Security Administration (a part of DOE)
OFE	Office of Fusion Energy
OFES	Office of Fusion Energy Sciences
OMB	Office of Management and Budget
ORMAK	Oak Ridge tokamak
ORNL	Oak Ridge National Laboratory
PLT	Princeton Large Torus

PPPL	Princeton University Plasma Physics Laboratory
SEAB	DOE Secretary of Energy Advisory Board
TAERF	Texas Atomic Energy Research Foundation
TFTR	Tokamak Fusion Test Reactor
TPX	Tokamak Physics Experiment

Contents

1 **Fusion Fundamentals** ... 1
Energy ... 1
Some Simple Nuclear Physics ... 2
Historical Origins of Fusion Research ... 3
Why Pursue Fusion? ... 5
Fusion Reactions ... 6
The Lawson Criterion ... 7
Heating ... 8
Other Key Technologies ... 10

2 **Fusion Concepts** ... 11
Magnetic Bottles ... 11
Inertial Confinement: Microexplosions ... 15
Other Concepts ... 15

3 **The Struggling Years: 1960s** ... 19
Magnetic Confinement Fusion ... 20
The Tokamak ... 22
Inertial Confinement Fusion ... 25

4 **The Glory Years: 1970s** ... 27
Tokamaks ... 28
Planning ... 29
Management ... 31
Tokamak Fusion Test Reactor ... 43
High-Density Systems: Pinches ... 45
Bumpy Torus ... 47
Open Systems: Magnetic Mirrors ... 48
The 1976 Magnetic Fusion Program Plan ... 49
US Department of Energy ... 52
Power Plant Designs ... 54
Surpassing the Lawson Ideal Ignition Temperature in a Tokamak ... 55
Fusion Power Associates ... 58

5 The Carter Plan vs. The Reagan Agenda: 1980–1985 ... 61
The Magnetic Fusion Energy Engineering Act of 1980 ... 61
The Pewitt Problem ... 63
Kintner Resigns ... 64
Mirrors and Bumpy Torus ... 68
Inertial Confinement ... 69
Management ... 69
ERAB Review ... 72
MFAC Strategy ... 73
Ansel Adams ... 74
William R. (Bill) Ellis ... 75
Tokamak Scaling ... 77
More on Inertial Confinement ... 77
Industrial Participation ... 79
Retrenchment ... 79
Trivelpiece Interview ... 81
Budgets and Reality ... 84

6 Successes and Disasters: 1985–1989 ... 87
Regrouping ... 88
Reagan–Gorbachev Summit Meeting ... 89
The Declaration of Energy Independence ... 90
Inertial Confinement Fusion Review ... 91
Halite–Centurion ... 94
The Compact Ignition Tokamak ... 94
The International Thermonuclear Experimental Reactor ... 95
Musical Chairs ... 95
ARIES ... 96
Congressional Testimony ... 96
The Arrival of Robert Hunter ... 98
The CIT Fiasco ... 99
More on Inertial Confinement Fusion ... 102
Fusion Policy at the Department of Energy ... 104

7 Hope for Resurgence: 1990–1995 ... 105
Input to Energy Secretary James D. Watkins ... 105
Fusion Policy Advisory Committee Review Begins ... 107
Bush–Gorbachev Summit Meeting ... 108
PBS Film: Fire from the Sun ... 108
FPAC Reports ... 109
Budget Cut ... 110
An Accelerated Fusion Power Development Plan ... 111
1991 National Energy Strategy ... 111
New Fusion Energy Advisory Committee ... 112
Happer at the Helm ... 113
ITER Moves Forward ... 115

Another Fusion Law ... 116
The 1992 and 1994 EPRI Fusion Reviews ... 117
Inertial Fusion and Beginnings of the NIF ... 120
Signs of Trouble ... 121

8 Financial Tsunami: 1995–1999 ... 125
Contract with America ... 125
The National Ignition Facility ... 126
The 1995 PCAST Fusion Review ... 127
Congress Takes Aim ... 128
Industry Groups Respond ... 130
Fusion Energy Mission Abandoned ... 131
TPX Reborn as KSTAR ... 136
1997 PCAST Energy Report ... 137
Pathways to Fusion Power Symposium ... 138
Farewell to TFTR ... 139
Fusion Community Attempts to Regroup ... 141
Congress Orders USA Out of ITER Collaboration ... 142
FPA Meeting: Cost-Effective Steps to Fusion Power ... 142
1998–1999 SEAB Review ... 143
1999 Fusion Snowmass Meeting ... 145
FESAC Comments on Fusion Program Priorities and Balance ... 145
OMB View ... 146
No Fusion Demonstration Power Plant by 2000 ... 146

9 The New Millennium: Science vs. Energy: 2000–2008 ... 149
National Energy Policy Development Group ... 150
National Academies Report on the Quality of Fusion Science ... 151
Burning Plasma Physics ... 152
ITER Rising? ... 154
Fifty Years of US Fusion Research ... 155
The High Average Power Laser (HAPL) Program ... 155
The 2002 Fusion Summer Study ... 156
Another Academies Fusion Review Panel ... 157
USA Rejoins ITER ... 158
The 35-Year Plan ... 159
Report of the Burning Plasma Assessment Committee ... 161
ITER vs. the US Domestic Fusion Program ... 163
Energy Policy Act of 2005 ... 165
More Proposed US Domestic Fusion Budget Cuts ... 166
Davies, Roberts, and Willis Retire ... 166
2007 Inertial Fusion Energy Workshop ... 167
Fifty Years of International Fusion Collaboration ... 169

10 The Obama Administration: 2009–2012 ... 171
New Appointments ... 171
Inertial Fusion Energy ... 173

NIF Begins Operation and Looks Beyond Ignition ... 174
Funding Improvement and Management Changes ... 175
ITER Changes ... 176
Magnetic Fusion Concepts Narrowing ... 177
Magneto-Inertial Fusion Approach Gets a Boost ... 178
Renewed Interest in Path to a Demonstration Power Plant ... 179
Inertial Fusion Energy Review ... 180
The FY 2013 US Fusion Program Dilemma ... 183
ANS Fusion Energy Division Comments on FY 2013 Budget Proposal ... 187
FESAC Launches Another Priorities Study ... 188
Congressional Actions ... 188
ITER Council Meets in Washington ... 191
Uncertainties ... 192

11 Applications ... 193
Electric Power ... 193
Hydrogen Production ... 194
Fusion–Fission Hybrid Reactors ... 194
Fuel for Nuclear Fission Reactors ... 195
Conversion of Nuclear Waste ... 195
Processing of Other Wastes ... 196
Desalination ... 196
Spin-Offs ... 197

12 Engineering Challenges ... 199
Materials ... 199
Tritium ... 201
Complexity ... 202
Maintenance ... 202
Cost ... 203

13 Energy ... 205
Primary Energy Consumption ... 205
Oil ... 206
Electricity ... 206
Coal ... 207
Natural Gas ... 207
Hydro ... 207
Nuclear ... 207
Renewables ... 208
Climate Change ... 208
Supply and Demand ... 209

14 Perspectives 2012 211
Charles C. Baker 211
Is the USA Serious About Fusion Energy? 211
N. Anne Davies 213
William R. Ellis 214
Richard D. Hazeltine 216
Fusion Science and Fusion Progress 216
Robert L. Hirsch 217
A Fusion Failure 217
B. Grant Logan 219
Taking a Longer, If Slower, Path to Fusion's Ultimate Potential 219
Robert L. McCrory 221
Perspectives on Inertial Fusion Energy: 2012 221
Dale M. Meade 222
Time for the Fusion Community to Focus on the Future 222
John H. Nuckolls 223
Inertial Fusion Energy: Super Lasers and Super Implosions 223
Richard F. Post 225
Magnetic Confinement Fusion Power Research: Thoughts After 60 Years of Involvement 225
Fred L. Ribe 226
Recollections and Perspective from 40 Years of Magnetic Fusion Research 226
John Sheffield 228
Ken Tomabechi 229
Alvin W. Trivelpiece 230

15 The Ultimate Energy Source? 233
Politics 233
Progress 235
Promise 237

Epilogue 239
Acknowledgements 244
Recommended Reading and Information Resources 244

References 245

Author Index 251

Subject Index 257

1 Fusion Fundamentals

> *The farther you can see into the past, the better will be your vision of the future.*
>
> –Winston Churchill

Energy

The term "energy" is defined by physicists as "the ability to do work." The laws of physics tell us that energy can neither be created nor destroyed, but only transformed from one form to another. In practical terms, Earth gets most of its energy from the Sun. The Sun is a large natural fusion energy source, sending its light and heat (forms of energy) into the surrounding solar system and into the universe beyond. Most stars are engaged in a similar process. The energy impinging on the Earth from the Sun is captured and stored in many ways but most importantly in living organisms. The Earth's extensive fossil fuel resources are the result of transformation and storage of the energy of living animals and plants over billions of years. Energy is also released from currently living organisms, most often by burning wood, but also by chemically transforming crops and other organic matter into burnable fuels.

In the early twentieth century, scientists discovered another source of stored energy on the planet: nuclear energy. This energy is released by converting mass from the nuclei of atoms into kinetic energy (energy of motion) of other nuclei and subatomic particles like neutrons. Einstein's famous formula $E=mc^2$ allows a precise calculation of this energy, which turns out to be millions of times more powerful than burning an equivalent mass of organic material. The fission (splitting) of heavy elements like uranium was the first form of nuclear energy found and harnessed. Fusion (combining) of the light elements, like hydrogen, was also discovered and found to be about eight times more energetic per unit mass converted than fission but also correspondingly more difficult to achieve in practice.

The history of evolution of life on Earth to the present day shows the important role that energy has played in that evolution. At first, humans used the energy stored

S.O. Dean, *Search for the Ultimate Energy Source: A History of the U.S. Fusion Energy Program*, Green Energy and Technology, DOI 10.1007/978-1-4614-6037-4_1,

in their own bodies from food to perform work. Then they discovered fire and, over many centuries, learned to use fire to improve their lives. Fire provided heat for homes, for cooking of food, and eventually for powering many new technologies, such as the steam engine. It was used in industry for chemical processing. Fossil fuels were burned for heat, for light, and, eventually, for electricity. The invention of electric power plants is just one example of how people learned to transform energy from one form into another and then to use the new form to do work more efficiently. The invention of the combustion engine, leading to the transportation systems in use today, is another example. In the mid-twentieth century, nuclear energy began to complement fossil fuels for the generation of electricity.

Some Simple Nuclear Physics

Fusion is a form of nuclear energy. Nuclear power plants currently provide about 20% of the world's electricity using fission—the process in which heavy nuclei of atoms are split into lighter ones. Fusion, on the other hand, as its name implies, is the process in which light nuclei are combined into heavier ones. It turns out, in both cases, that mass is converted into energy in accordance with Einstein's famous formula $E=mc^2$, where c is the velocity of light and m is the amount of mass that "disappears" in the conversion process. The velocity of light being a very large number, it takes only a very small amount of mass conversion to produce a very large amount of energy.

Nuclei are composed of protons (which have an electrical charge) and neutrons (which have no charge). Nuclei are stable when they have just the right combination of protons and neutrons in the nucleus. Upset this balance in the nucleus of one element, and it tends to change into the nucleus of another element. It turns out that at both the light and heavy element ends of the periodic table of the elements, these changes result in the conversion of mass into energy.

It is much more difficult to combine nuclei than to split them. Practically speaking, it is relatively easy to split (i.e., fission) a heavy element like uranium and relatively hard to combine (i.e., fuse) light elements like hydrogen. This is because a heavy element can be caused to fission by inserting an additional uncharged neutron into its nucleus, thereby destabilizing it; whereas to fuse two nuclei, one must overcome the natural repulsive force due to the electrical charge of the protons in their respective nuclei (so-called Coulomb force). This latter force is very strong and increases rapidly the closer the nuclei approach each other. It is for this reason that fusion is only observed in nature in the Sun and stars at extremely high temperature (millions of degrees), i.e., when the nuclei are approaching each other at very high speed.

Many, if not most, of the basic approaches to achieving fusion were postulated in rudimentary form during the 1950s based on well-known principles of electromagnetic theory and nuclear physics [2]. Nuclei carry positive charge and can therefore be guided by a magnetic force field. Since the repulsive (Coulomb) force between two nuclei increases the larger the charge on the nuclei, fusion would be most easily achieved, it was reasoned, between singly charged isotopes of hydrogen. Even then, however, two nuclei approaching each other are about a thousand times more likely to be scattered than to fuse, even when traveling toward each other at high speed. Consequently, some means is needed to confine the nuclei for a sufficient time to allow many attempted collisions. Magnetic bottles, of various shapes, seemed the ideal solution (see Chap. 2).

Historical Origins of Fusion Research

The origins of fusion physics can be traced to the origins of nuclear physics and are a natural evolution of it. Sir Ernest Rutherford received the Nobel Prize in 1908 for his seminal work on the theory of atomic structure and for showing how radioactive elements transform into other elements of the periodic table. In 1919, Rutherford also performed experiments showing how heavier elements could be produced by the collision of lighter elements—a process later to be called "fusion." In the same year, British physicist Francis Aston, using a mass spectrometer he had invented (and for which he later also received a Nobel Prize), demonstrated the existence of different "isotopes" of the same element (i.e., atoms of the same element that have different atomic weights) and discovered the then-astonishing fact that the mass of a helium nucleus was less than the sum of the hydrogen nuclei of which it was composed. In 1920, Sir Arthur Eddington, in a speech to the British Academy for the Advancement of Science, suggested that the "fusion" of light elements, starting with hydrogen to helium, was the source of energy in the Sun and stars.

The lightest element in the periodic table, hydrogen, consists of a single proton and has two isotopes, deuterium (with a proton and a neutron in the nucleus) and tritium (with a proton and two neutrons in the nucleus), as shown in Fig. 1.1. Deuterium was first identified in 1932, tritium in 1934, though the existence of both had been predicted earlier. During the 1930s, scientists around the world were actively investigating the makeup of the periodic table and the reactions that might explain the energy processes of the Sun and stars. In 1934, Rutherford and his colleagues demonstrated the fusion of deuterium and deuterium to form helium using a Cockcroft–Walton particle accelerator.

A 1936 review article by Hans Bethe in the journal Reviews of Modern Physics led to the first recorded interest in building a fusion experiment in the USA in 1938, that of Kantrowitz and Jacobs at the Langley lab of the National Advisory Committee for Aeronautics (NACA, the predecessor agency of NASA). They built a simple torus (donut-shaped) vessel with magnet coils wrapped around it to produce a "magnetic bottle" and introduced about 150 W of power from a radio transmitter, hoping to heat the hydrogen gas to a million degrees. The experiment failed to produce the desired result and was abandoned. In 1939, Australian physicist Peter Thonneman conceptualized a fusion reactor. He later played a key role in the British fusion program [3].

Scientists in the UK and the USA working on the atomic bomb during World War II were already thinking beyond fission to fusion. In 1946, two British physicists,

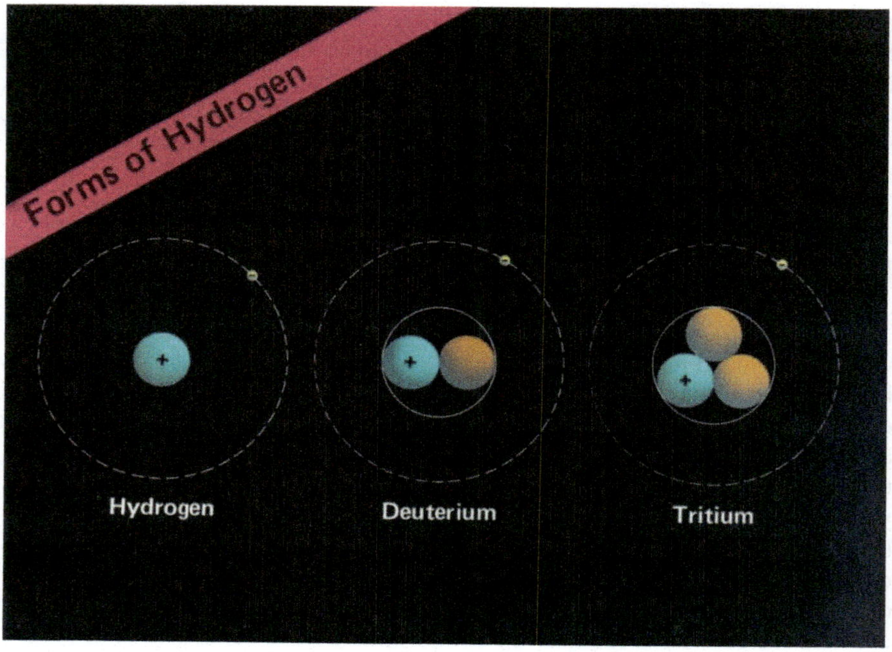

Fig. 1.1 Isotopes of hydrogen. An ordinary hydrogen nucleus contains one positively charged proton. The two "heavy isotopes" of hydrogen are deuterium (which also contains one neutron in the nucleus) and tritium (which contains two neutrons). Also shown is the small negatively charged electron that orbits the nucleus to form hydrogen atoms. In hydrogen gas at high temperature, the electron is stripped from the atom, and the resulting collection of freely moving electrons and nuclei is called a plasma

Sir George Thompson and Moses Blackman, filed a secret patent application for a donut-shaped, current-driven "pinch" fusion device they had designed at Imperial College. In 1947, two doctoral students at Imperial College, Stan Cousins and Alan Ware, built and operated a small magnetic "pinch" experiment there (see Chap. 2 for discussion of the "pinch" and other concepts.) One of those active in the UK fusion program in those early days was Jim Tuck, who had worked at Los Alamos during the war, returned to the UK, and then returned to Los Alamos to work on the H-bomb. Tuck, known for his sense of humor, built a pinch device in 1952 at Los Alamos he called the "Perhapsatron" (Fig. 1.2) because, he said, "perhaps it will work and perhaps it will not." Others joked that perhaps it should be called the "Impossibilitron" [1].

Although during the 1940s scientists in the US nuclear weapons laboratories informally discussed fusion from time to time, largely in the context of how to build a hydrogen bomb, the real stimulus for controlled fusion research is usually attributed to a front-page article in the New York Times, March 25, 1951, reporting that Argentina claimed to have a working fusion power plant. Though the claim ultimately proved untrue, it gained the attention of scientists and politicians in the USA,

Fig. 1.2 Perhapsatron—the first US fusion experiment at Los Alamos—consisted of a simple donut-shaped vessel filled with deuterium gas surrounded by magnets (here shown half cut away). A current was then induced in the gas, forming a plasma and heating it (Courtesy Glen Wurden, Los Alamos National Laboratory)

UK, and USSR, and serious efforts to investigate the possibilities were launched. Soon thereafter in the USA, experiments were under way at Los Alamos under Jim Tuck, at Princeton University under Lyman Spitzer, and at the University of California Radiation Laboratory (now Lawrence Livermore National Laboratory) under Edward Teller, Herb York, and Richard F. Post. The US program was highly classified; the British and Soviet programs perhaps a little less so [1, 4].

In the Soviet Union, it is believed that Oleg Lavrentiev was the first to call attention to fusion, in letters he wrote to the government in 1949 and 1950 [5]. These letters in turn aroused the interest of Soviet scientists Igor Tamm and Andrei Sakharov. In 1951, the government formally launched a fusion program under the direction of Igor Kurchatov, director of the Institute of Atomic Energy in Moscow. He set up an experimental program under the direction of Lev Artsimovich and a theoretical effort under the direction of Mikhail Leontovich. Other institutes were later brought into the effort from Kharkov, Leningrad, and Sukhumi.

Why Pursue Fusion?

The many reasons why scientists and engineers pursue fusion for energy include the following:

- It will provide a universally available, essentially inexhaustible, efficient fuel resource. The primary fuel, deuterium, can be obtained inexpensively from water. Tritium fuel can be produced from lithium, an abundant resource. Fusion reactions release about eight times more energy than fission reactions and about a million times more energy than burning fossil fuels, per unit weight of fuel consumed. And fusion is a more concentrated energy source than solar or wind.

- It will provide an option to meet growing global energy demand. Fusion, fission, and various non-hydro renewables are the only known long-term alternatives to continued reliance on fossil fuels. Fossil fuels, especially oil, are a limited resource, not universally available, and cause environmental damage. Future public acceptance of fission and fossil is uncertain. And the future economics and ability to meet demand by the non-hydro renewables is also uncertain.
- It will provide an energy source with attractive safety and environmental characteristics. Compared to fossil energy, fusion has no chemical combustion products, no greenhouse gas emissions, no large-scale fuel mining and transportation requirements, and no large volume of waste products. Compared to fission, fusion has no possibility of "criticality" or reaction "runaway" leading to "meltdown," shorter-lived less biologically hazardous radioactivity, and no weapons grade fissionable material. Compared to solar and other non-hydro renewables, fusion has less mining and use of toxic materials, no susceptibility or dependence on cloud cover or weather, no requirement for energy storage, is more compatible with large-scale generation, and has greatly reduced land requirements.
- It will provide new technologies for a variety of applications (see Chap. 11). Fusion is at the cutting edge of science and technology. It has potential applications other than for electric power, including production of hydrogen as a fuel for transportation or for fuel cells, desalination of water, destruction of radioactive wastes, production of fuel for fission reactors, production of special isotopes, and food sterilization. Fusion science and technology have many near-term ("spinoff") commercial applications, including destruction of toxic wastes, fabrication of micro-miniature integrated circuits, deposition of anticorrosion coatings, applications of superconductivity, ultraviolet light sources, laser applications in industry and medicine, X-ray lithography, welding, printing of polymer films, production of high-performance ceramics, surface cleaning of materials, and precision optics and diagnostic equipment.

Fusion Reactions

Overcoming the repulsive Coulomb force between charged nuclei is easiest for those nuclei having the lowest charge. Isotopes of hydrogen have charge one, hence are the isotopes of choice for fusion, although some researchers also consider the fusion of deuterons with helium (charge of two) or protons with boron (charge of five). Overcoming the Coulomb force also requires that the two nuclei be approaching each other at high speed (in a gas, this implies high temperature).

Measurements of the probability of two nuclei fusing show that the peak probability of fusing deuterons and tritons is about six times higher than the fusion of two deuterons and occurs at about a fivefold lower temperature, as shown in Fig. 1.3. Fusion among elements of higher than charge one occurs with even lower probability and requires even higher temperature. For these reasons, fusion of deuterons with tritons has been the reaction of choice to date.

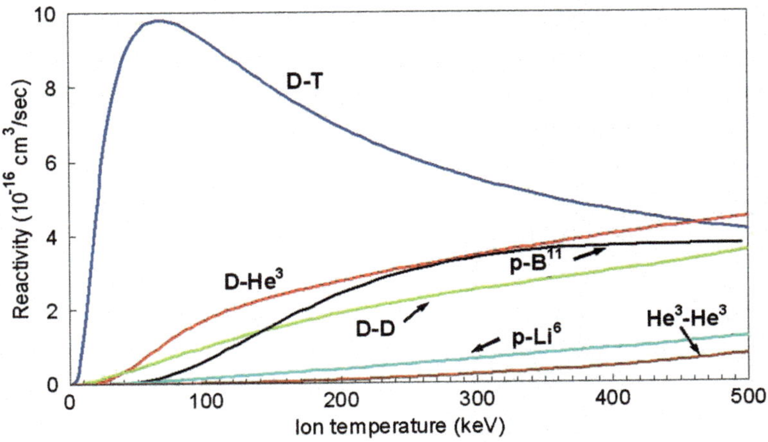

Fig. 1.3 The reactivity (probability) of fusion reactions is much higher for deuterium–tritium (DT) and occurs at lower temperature than for other fusion reactions such as DD

When a deuterium nucleus (deuteron) fuses with a tritium nucleus (triton), it produces a helium nucleus and an extra neutron, as shown in Fig. 1.4. The combined energy of the helium nucleus and the neutron is about 1,000 times larger than the energy required to overcome the repulsive Coulomb force between the deuteron and the triton. In this sense, a fusion reactor is an "energy amplifier." It requires energy to initiate the reaction but amplifies that energy about a thousandfold.

The Lawson Criterion

The energy released in one fusion reaction is not of practical interest. To achieve a significant amount of energy, many millions of reactions must be continuously produced. This requires a sufficiently high density of nuclei (easily achieved in an ionized gas called a plasma) at a sufficiently high temperature (somewhat more difficult to achieve) and "contained" for a sufficiently long time (quite difficult to achieve) to result in a specific amount of energy of interest to be released.

In a 1957 landmark paper, British physicist John D. Lawson calculated the density, temperature, and "confinement time" required for a fusion power plant [6]. He showed there was a minimum temperature that must be exceeded (called the ideal ignition temperature) but that a trade-off existed between the required density (n) and confinement time (τ): higher density required lower confinement time and vice versa. These requirements (on temperature, T, and the product of density and time, $n\tau$) came to be known as "the Lawson criterion" and constituted a scientific "Holy Grail" for fusion researchers (sometimes called proof of "scientific feasibility"). This process is shown schematically in Fig. 1.5.

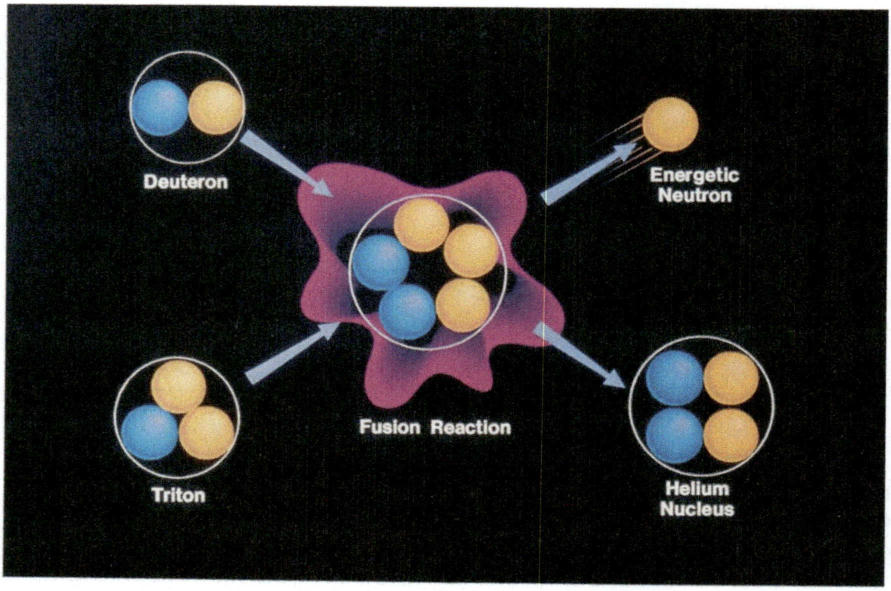

Fig. 1.4 Fusion of a deuteron with a triton to produce a fast-moving helium nucleus and an energetic neutron. The kinetic energy (energy of motion) of the products (helium nucleus and neutron) can be recaptured by slowing them down in an energy-absorbing material (often called a "moderator" or "blanket")

Heating

Starting with a gas of hydrogen at room temperature, the simplest way to begin heating it toward fusion temperatures is to pass an electric current through it, creating an electrical discharge. This is a sort of super fluorescent light or like a lightning bolt. This current creates a "plasma" or ionized gas, in which some or all of the electrons are stripped from the hydrogen atoms, resulting in a collection of hydrogen nuclei and electrons. This plasma has an electrical resistance to the current which heats the plasma in the same way that current through a wire heats the wire or lights up the filament in a light bulb. However, as the temperature of the plasma rises, its electrical resistance lowers, and further temperature increases become more difficult. The result is that there is a limit to how high a temperature can be achieved this way, even with very large currents. Unfortunately, the limiting temperature is less than the ideal ignition temperature required for fusion. This technique, called *ohmic heating*, was among the earliest tried in experiments and is still often used to provide a starting point for subsequent "auxiliary" heating to higher temperatures.

Another technique, *compression*, is also effective in heating plasma, once a plasma has been formed. This usually takes the form of surrounding the plasma with a magnetic field, either formed internally by the ohmic current or by an

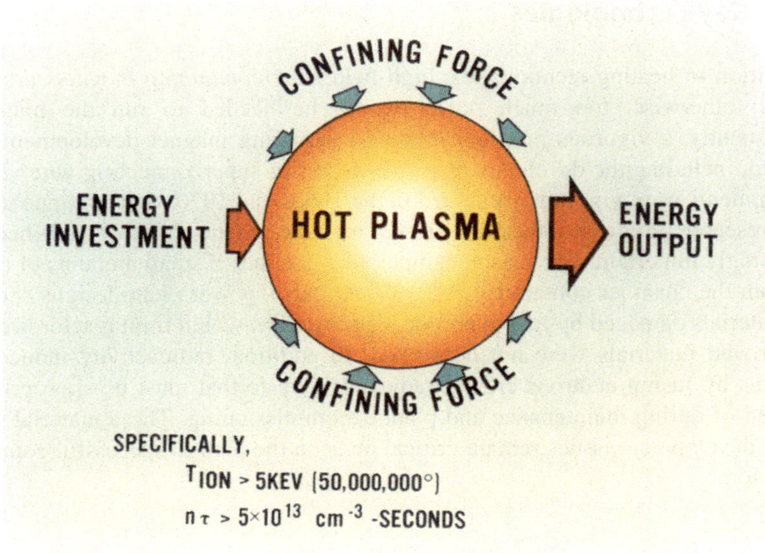

Fig. 1.5 Schematic showing energy investment in producing (deuterium–tritium) hot plasma produces larger energy output if its temperature and density-confinement time product exceed minimum values (as calculated by Lawson)

external field provided by magnets. When this magnetic field is rising in time during the experiment, it exerts a force on the electrically charged plasma, thus compressing it (raising its density and further increasing its temperature). Compression has the advantage that very high density and temperature can be achieved this way. Though very effective, compression is most compatible with fusion concepts that are inherently pulsed, whereas many fusion researchers prefer technologies that are more compatible with concepts that may operate in steady state at lower density.

A third technique that has found widespread use is the injection of *beams of particles* from an accelerator. Usually, these beams start out as charged hydrogen nuclei that are first accelerated, then charge-neutralized and injected into a previously formed plasma. These *neutral beams* can have energies far in excess of the energy corresponding to the ideal ignition temperature. They then may be trapped and confined themselves or heat the existing plasma by collisions. This technique was not available in the early history of fusion research but now is widely used in experiments and in conceptual fusion power plant designs.

A fourth technique, gaining favor rapidly, is to couple power from a powerful *radio-frequency source* directly into the plasma via a resonant absorption process. There are many frequency possibilities and many absorption mechanisms being studied. This technique is viewed as an especially elegant technology because of its sophistication and compatibility with steady-state operation.

Other Key Technologies

In addition to heating technologies, high-field *superconducting magnets* are also needed; otherwise, too much power would be needed to run the magnets. Consequently, a vigorous program in superconducting magnet development was initiated, including the development of new forms of superconducting wire. These developments were vigorously pursued in the 1960s and 1970s and continue today.

As research progressed, the *properties of materials* in the chamber walls became increasingly important. The plasma temperature declines if small amounts of material from the chamber contaminate the plasma. Also, power plant designs showed that materials damaged by fusion products hitting them would limit reactor lifetime if improved materials were not developed. In addition, radioactivity induced in materials by fusion neutrons creates radioactive waste that must be appropriately disposed of during maintenance and plant decommissioning. These material technology development issues remain critical ones on the path to successful commercial fusion.

Fusion Concepts 2

Good ideas are not adopted automatically.
They must be driven into practice with courageous patience.

– Admiral Hyman G. Rickover

Magnetic Bottles

Early efforts to produce and control fusion reactions were based on then well-known principles of electromagnetic theory. A current passing through a gas was known to strip electrons from the gas atoms (ionization), to raise its temperature, and to produce a magnetic field surrounding the current. Raising the current increased the degree of ionization, the temperature, and the magnetic field strength. The magnetic field exerts a confining force on the column of ionized gas (dubbed "plasma" in a 1928 paper by Irving Langmuir), and as the current and magnetic field is raised, the column of plasma is compressed, raising its density and further raising its temperature. This was known as the "pinch effect" and was the basis of most of the early attempts to produce fusion conditions in the laboratory. The "pinch effect" had been predicted in 1934 by W. H. Bennett and, independently, in 1937 by Lewi Tonks, but little subsequent effort was devoted to pinch plasma properties in the 1930s. Later, pinch devices were fashioned into what came to be called "magnetic bottles" for the plasma. In the 1950s, some of these "magnetic pinch" devices studied for fusion were linear in geometry, and some were donut-shaped (toroidal). They went by a variety of sometimes-colorful names: Perhapsatron and Columbus at Los Alamos and Zeta in the UK [2].

Pinches had serious problems, however. It was observed that, as the plasma column was pinched, the plasma twisted and moved in "unstable" fashion and quickly hit the walls of the chamber. A variety of such "instabilities" were observed in the various pinch configurations, receiving names such as "kink instability" and "sausage instability." Figure 2.1 illustrates a linear pinch configuration (a) before any instabilities develop and (b) with a kink instability which disrupts the discharge. Figure 2.2 illustrates a toroidal pinch configuration. Much of the first two decades

S.O. Dean, *Search for the Ultimate Energy Source: A History of the U.S. Fusion Energy Program*, Green Energy and Technology, DOI 10.1007/978-1-4614-6037-4_2,

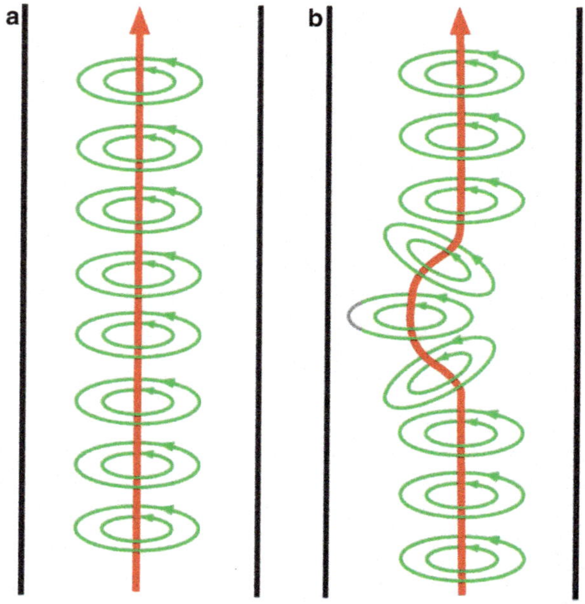

Fig. 2.1 Linear pinch. A current (*red line*) is driven through a gas in a cylindrical chamber, creating a plasma. Currents have magnetic lines of force surrounding them. If the current is increased, the magnetic force becomes stronger, compressing and heating the plasma (**a**). But if a slight asymmetry develops, the force becomes unbalanced, resulting in a "kink instability" (**b**), which drives the plasma into the wall of the chamber where it is cooled and lost (US Atomic Energy Commission; Amasa Bishop, *Project Sherwood*, Addison-Wesley Publishing Company, 1958)

of fusion research was devoted to developing an understanding of these and related instabilities, all belonging to a class to be known as Magnetohydrodynamic (MHD) instabilities or macroscopic instabilities.

It was early recognized that plasma would rapidly leak out the ends of a linear magnetic bottle unless something was done to "plug" the ends. One "solution" already mentioned is the toroidal configuration (which has no "ends"). Another "solution" emerged in the form of strengthening the magnetic field at either ends of the linear configuration. This geometry came to be known as the "magnetic mirror" configuration (Fig. 2.3) and was championed in the USA by Richard F. Post and colleagues at the University of California's Livermore Laboratory [7]. While magnetic mirrors have been largely written-off as fusion power plants, due to the still rapid loss of plasma out the ends while controlling radial transport, there is still belief in some quarters that they could experience a resurgence [8, 9]. If so, they would have attractive power plant features, including simple geometry for maintenance and the potential for more efficient energy conversion, e.g., avoidance of the conventional heat-to-steam-to-electricity cycle.

Another early variant of the toroidal magnetic bottle was invented by Princeton University astrophysicist Lyman Spitzer while riding the ski lifts in Aspen. He asked himself how one might contain on Earth a plasma similar to that existing in stars. He envisioned a magnetic configuration he called the "stellarator" since it was designed to contain a man-made equivalent of a star on Earth. It was similar in some respects to the toroidal pinch configuration but differed in that external magnets produced the primary confining magnetic field rather than a current in the plasma. Even in this case, however, the plasma had many surprises in store for the researchers.

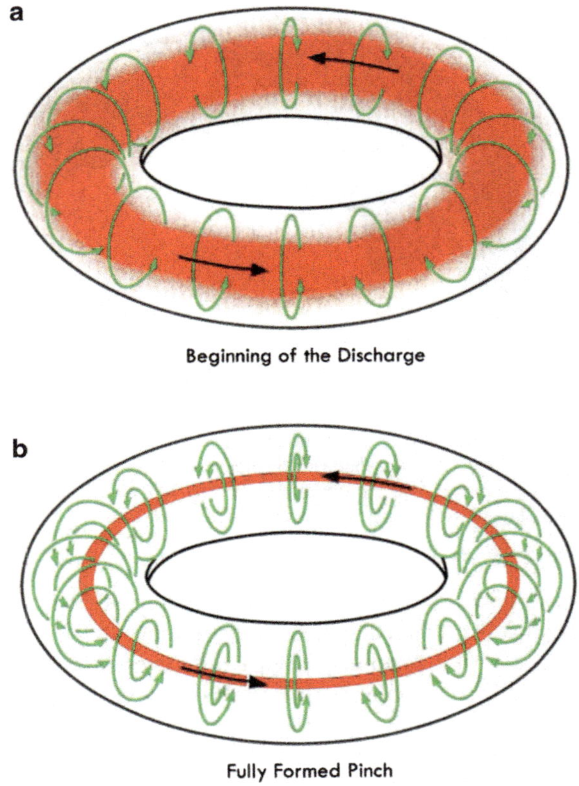

Fig. 2.2 Toroidal pinch. Instabilities (not shown) similar in nature to those seen in linear pinches are present also in toroidal configurations (US Atomic Energy Commission; Amasa Bishop, *Project Sherwood*, Addison-Wesley Publishing Company, 1958)

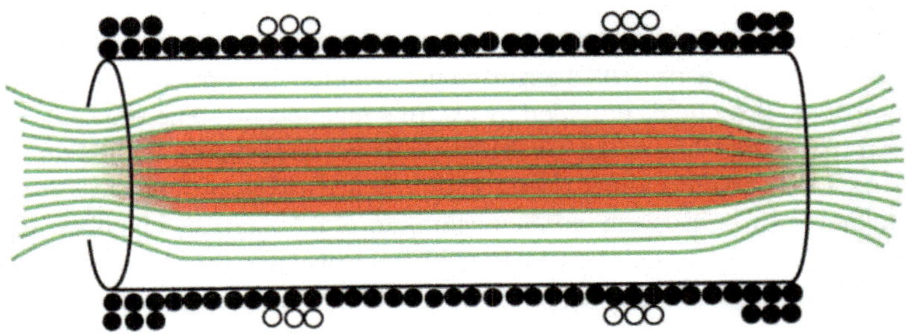

Fig. 2.3 Magnetic mirror configuration: stronger magnetic field strength at ends hampers loss of fusion plasma (US Atomic Energy Commission; Amasa Bishop, *Project Sherwood*, Addison-Wesley Publishing Company, 1958)

These, more subtle, types of plasma loss mechanisms in the stellarator geometry came to be known as "microinstabilities" and are still the subject of active research. The stellarator is still a promising configuration for fusion plasma confinement, although the magnets needed to provide the magnetic field properties necessary to suppress instabilities have become increasingly complex to manufacture. Figure 2.4

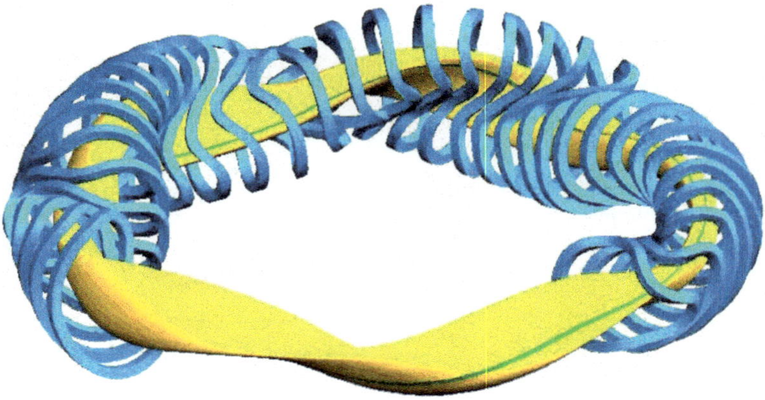

Fig. 2.4 Schematic of W7X stellarator plasma and magnets under construction in Greifswald, Germany. The complex shape of the plasma and magnet coils are believed necessary to provide the desired stability and confinement of the plasma in a stellarator (Max-Planck-Institut fur Plasmaphysik (IPP) in Greifswald)

shows a schematic of a modern stellarator with its complex magnet coils surrounding what is predicted to be a stable plasma.

As part of the Atoms for Peace initiatives of the late 1950s, and as scientists realized the extreme complexity of plasma behavior in fusion experiments, the USA, the UK, and the Soviet Union agreed to remove the veil of secrecy that had surrounded their fusion research efforts and to present their research programs at the Second United Nations Geneva Conference on the Peaceful Uses of Atomic Energy, in Geneva, in 1958. Thus began a spirit of friendly competition and cooperation among fusion scientists worldwide that has lasted to the present day. At the Geneva conference, the Soviets described experiments in which the toroidal pinch geometry was supplemented by fields provided by external magnets. In some ways, the geometry resembled a marrying of the pinch and stellarator ideas being studied separately elsewhere. One version of this configuration that evolved during the 1960s was called tokamak (from Russian words meaning toroidal magnetic chamber). By the late 1960s, this configuration showed dramatic improvement in confining the plasma compared to other geometries. A worldwide shift to this configuration began after the 1968 fusion conference sponsored by the International Atomic Energy Agency (IAEA) in Novosibirsk. The tokamak configuration, shown schematically in Fig. 2.5, now dominates world fusion research.

During the 1970s and 1980s, tokamaks capable of creating "near-breakeven" conditions (in which the fusion energy released approximately equals the energy content of the plasma) were constructed and successfully operated in the USA, Europe, and Japan. The experience gained from these and other tokamaks around the world has led to initiation of the International Thermonuclear Experimental Reactor (ITER) being constructed in France as a joint venture of the European Union and six other country partners and currently scheduled to begin operation about 2020.

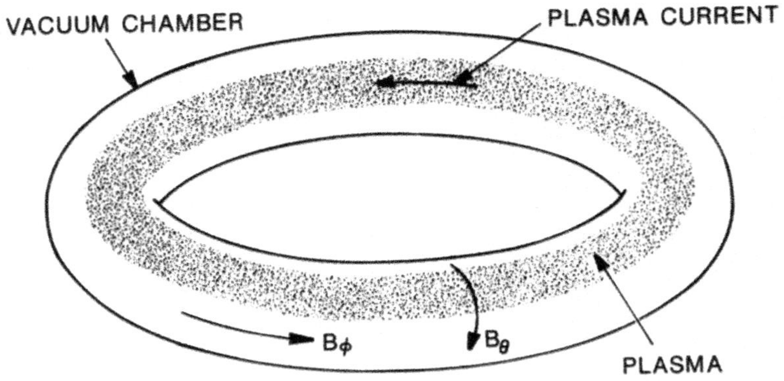

Fig. 2.5 Schematic of tokamak magnetic configuration. Plasma current produces the magnetic field B_θ shown surrounding the plasma, as in the pinch configuration, and magnet coils (not shown) add an additional magnetic field B_ϕ around the torus

Inertial Confinement: Microexplosions

The invention of the laser in 1960 gave rise to a whole new approach to fusion called "inertial confinement fusion." The hydrogen bomb showed that fusion could be initiated by a sufficiently strong compressive force exerted on a small amount of fusion fuel. In the case of the bomb, this force was provided by a fission-based atomic bomb surrounding the fusion fuel. In the 1960s, scientists in and outside the weapons laboratories began to speculate on whether a fusion reaction of practical interest could be initiated by focusing a high-power laser on a small capsule containing fusion fuel. A schematic of this process is shown in Fig. 2.6. Its pioneers compiled a history in 2007 [10]. The early studies showed that, while inertial fusion was theoretically possible, lasers far beyond those available at the time would be required.

Larger and larger lasers were built during the 1970s and 1980s, culminating in the National Ignition Facility (NIF), located at the Lawrence Livermore National Laboratory in California. This 192-beam laser, which began operation in mid-2009, is designed to produce a net output of fusion energy compared to the laser energy required to initiate the fusion reaction. A similar laser is under construction in France, scheduled to be completed in 2014.

In addition to lasers, other forms of pulsed power (ion beams, z-pinch) are potential "drivers" for compressing fusion pellets and are being pursued, e.g., at Lawrence Berkeley National Laboratory, Sandia National Laboratories, and elsewhere.

Other Concepts

The plasma "likes" the "closed" magnetic field geometry of stellarator and tokamak configurations, which provide relatively good plasma confinement. Unfortunately, this physics advantage comes with an engineering disadvantage. The donut-shaped

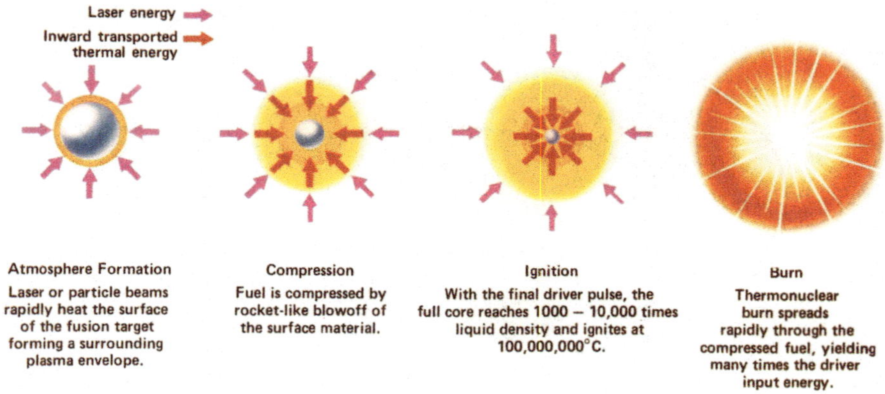

Fig. 2.6 Schematic of inertial confinement fusion (ICF) using a laser or other driver. Energy impinges on a capsule containing fusion fuel, and the resulting compression produces the high density and temperature required to initiate fusion reactions (Lawrence Livermore National Laboratory)

toroidal vessel has a hole in the center, which makes access difficult for maintenance. Neutrons from the fusion reactions also converge inwards, so that materials on the inner section of the vessel may have to be replaced frequently. The ideal geometry for a fusion power plant, from an engineering and maintenance viewpoint, would be one in which the vessel is cylindrical (or perhaps spherical) with the plasma at the center so that all maintenance could be performed from the outer periphery. Several configurations have been identified and studied with this property, including the field reversed concept (FRC) (Fig. 2.7) and the Spheromak. Several investigators are pursuing fusion concepts in which, using an FRC plasma as a starting point, the FRC is manipulated in various ways, e.g., compressed, sustained by beams, translated, and confined.

One of the most advanced concepts incorporating an FRC plasma is magnetized target fusion (MTF), or magneto-inertial fusion (MIF), being pursued at the Los Alamos National Laboratory, Sandia National Laboratories, General Fusion, and elsewhere. In this concept, an FRC plasma is formed either in situ or translated in space to a cylindrical chamber that is surrounded by a conducting shell that is then imploded onto the plasma, compressing and heating it (Fig. 2.8). Of the MTF concept, Freidberg notes [11] that it "offers a true alternative to conventional magnetic fusion concepts, and in this context is certainly worth examining as a potential source of fusion energy." Other variations, similar to the FRC, include the spheromak [12] and the use of an FRC to confine a beam of colliding ions [13] (See also [14]). The private company TriAlpha in California is pursuing the latter.

The company EMC2, founded by Robert W. Bussard, is currently exploring another concept, called inertial electrostatic confinement (IEC). IEC was originally conceived by Philo T. Farnsworth, the inventor of television, and then pursued by Robert L. Hirsch. Later, Bussard would become a major advocate of IEC. In addition to Bussard, aspects of IEC have also been explored by scientists, including Nicholas Krall (consultant), George Miley (University of Illinois), and Gerald

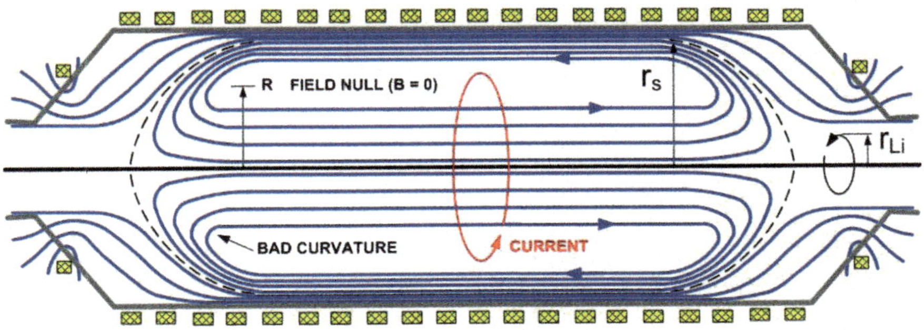

Fig. 2.7 Schematic of field reversed concept (FRC). This configuration combines the advantage of having closed magnetic field lines (as in a torus) with the advantage of having a simpler linear cylindrical vessel (as in a magnetic mirror) (Francis F. Chen, *An Indispensable Truth*, Springer)

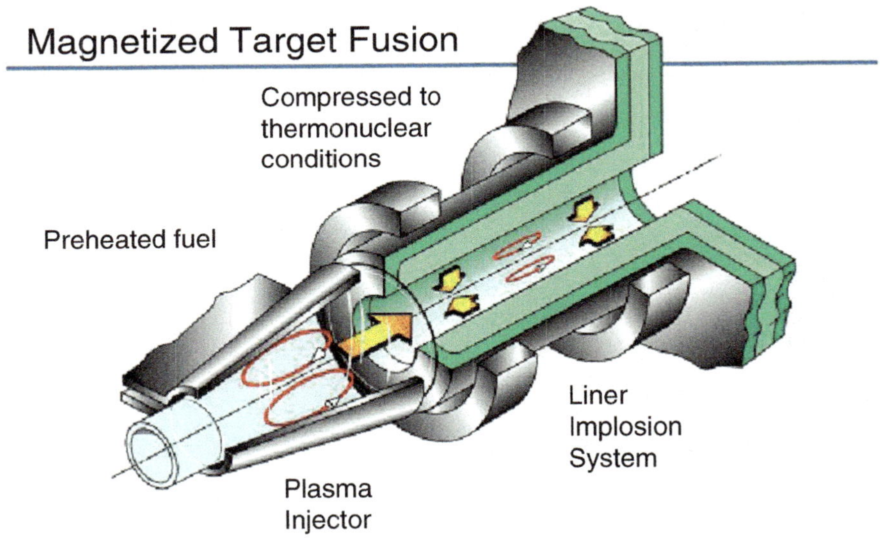

Fig. 2.8 In magnetized target fusion (MTF), a plasma is preheated slightly and injected into a chamber; the walls of the chamber ("liner") are then imploded inwards, compressing the plasma to fusion conditions (Los Alamos National Laboratory)

Kulcinski (University of Wisconsin). IEC has been shown to be a good point source of fusion neutrons for some applications, but its scaling to fusion reactor conditions is still uncertain.

While each of these "other" concepts has attractive power plant characteristics, a "breakthrough" demonstration equivalent to the tokamak has eluded the advocates of these concepts, at least in the small-scale experimental facilities available to them. It remains to be seen whether the attractive features of these concepts can be validated in future facilities. Unfortunately, with limited resources, government managers of the world fusion programs have felt constrained to restrict funding for these "alternate" (to the tokamak) concepts.

The Struggling Years: 1960s

3

The only thing worse than failure is not to try at all.

–NASA Administrator James Beggs

As fusion research progressed, countries other than the USA, UK, and USSR initiated substantial fusion research efforts and began making important contributions to what is a vigorous international collaborative effort. These included Germany, France, Japan, Republic of Korea, China, India, and others. The European Union (EU) is now coordinating fusion research among all the EU countries, and the United Nation's International Atomic Energy Agency (IAEA) has continued its world fusion coordination activities, begun in 1958, through its biennial conferences, technical working groups, and the International Fusion Research Council (IFRC).

After reading Bishop's book in the summer of 1958, I resolved to make my career in fusion. Unfortunately, I found that the field was so new that almost no universities offered graduate education in the newly recognized subject of plasma physics (the science that underpins fusion plasma behavior) or in fusion engineering. One exception I found was the Nuclear Engineering Department at the Massachusetts Institute of Technology (MIT). It offered a fusion option (as contrasted with the more common fission option) as a major, under Professor David J. Rose. However, as a physics major at Boston College, I did not seem to have the courses necessary to qualify to enter a nuclear engineering department, most of whose students came with a chemical engineering undergraduate background. To offset this disadvantage, I applied for and was accepted to take a heat engineering course in the summer of 1959 at the Rensselaer Polytechnic Institute in Troy, NY, taught largely by nuclear engineers from the Knolls Atomic Power Laboratory (KAPL). The Knolls' engineers were designing fission reactors for Admiral Rickover's emerging nuclear navy. In the fall of 1959, I applied for and received an Atomic Energy Commission Graduate Fellowship. Armed with that Fellowship, I applied and was accepted to the MIT Department of Nuclear Engineering in the fall of 1960.

S.O. Dean, *Search for the Ultimate Energy Source: A History of the U.S. Fusion Energy Program*, Green Energy and Technology, DOI 10.1007/978-1-4614-6037-4_3,

While studying at MIT with Professor Rose and others, I learned a lot about fission, fusion, and nuclear science and technology. In lieu of doing a Master's thesis, the department offered, and I chose, the opportunity to spend the Spring (1962) semester working on various projects at the Oak Ridge National Laboratory in Tennessee. One of the projects I worked on was to measure the vacuum pumping speeds on the new magnetic mirror fusion device there called DCX-II. The contacts I made there allowed me to successfully apply for a job to join the AEC's Controlled Thermonuclear Research (CTR) management office in Washington, DC, after graduating from MIT in June 1962.

Magnetic Confinement Fusion

The CTR branch was part of the AEC's Division of Research, headed by Dr. Paul McDaniel. The branch was headed by Dr. Arthur E. Ruark, a crusty old professorial type who, it turned out, had a very small staff and a soft spot for helping to launch the careers of young people. When I arrived, the only person on Ruarks's staff was William C. Gough. Soon thereafter, we were joined by Dr. Hilliard Roderick. As I was preparing for my first visit to see the fusion programs at the Lawrence Livermore lab, Roderick advised me not to meet with either Harold Furth or Sterling Colgate there. They were two of Edward Teller's "whiz kids" and would simply look down their nose and insult me, he said. I did, however, meet with them and found them both respectful and entertaining.

My first task in the CTR branch was to review and process a contract renewal proposal from Harold Grad's mathematical theory group from New York University. Ruark always referred to that group as "the conscience of the community" because they were more rigorous (but less practical) in their mathematical treatments of plasma theory than the bulk of the theorists working on fusion problems. This appealed greatly to Ruark. I soon discovered, however, that Grad's approach, while important to the emerging field of plasma science, did not often result in solutions to equations that could be compared to experiment, while other theorists, of whom Marshall Rosenbluth was the acknowledged leader, were able to provide more guidance to experiments, even though they had to make numerous assumptions and simplifications to get solutions to the equations. The "confrontations" between these two schools of theorists provided many entertaining and educational dialogues at fusion meetings throughout the 1960s. This "conflict" still exists today between those who rely primarily on "semiempirical" models of plasma behavior and those who advocate concentrating on seeking to develop a "predictive capability from first principles."

One of the more interesting features of the fusion program I entered in 1962 was a privately funded group at General Atomics (GA) in San Diego (a subsidiary of General Dynamics at the time). The company had been set up in 1955 to pursue research in the nuclear fission field but, in 1956, secured funding from the Texas Atomic Energy Research Foundation (TAERF) for a jointly funded ($5 million each), 4-year (which was subsequently extended) fusion effort. Frederic ("Freddie") De Hoffmann, who hired Edward Creutz of the Carnegie Institute of Technology as research director, headed the company. Creutz brought in some of the top fusion scientists of the time (both theorists and experimentalists), including Marshall Rosenbluth and Donald Kerst. They, in turn, attracted other talent, including William Drummond, Nicholas Krall, Norman Rostoker, Bruno Coppi, Ken Fowler, Charles Wharton, and a talented young scientist from Japan, Tihiro Ohkawa. I visited GA several times over the next few years and was always impressed with the team. They were very focused but still explored a variety of magnetic configurations and were always inventing new variations.

During the 1960s, things were not going well in US experiments. Stellarators at Princeton Plasma Physics Laboratory (PPPL) were observing loss rates of plasma more rapid than expected, and the reasons were not understood (though there were many theories). The other major approach, the magnetic mirror at the Lawrence Livermore Laboratory (LLL), likewise was struggling, with the exception of the so-called 2X mirror there, which used plasma compression. A third (and unique) approach, Astron (also at LLL), was also struggling [15]. Similar problems were besetting the experiments in other countries.

The US fusion program in the 1960s was overseen in the US Congress by the Joint (House–Senate) Committee on Atomic Energy (JCAE). The JCAE was growing impatient with the slow pace of fusion development. At the JCAE hearings on the AEC's FY 1965 budget in the spring of 1964, Chairman John Pastore said, "how long do you have to beat a dead horse over the head to know that he is dead?" The JCAE demanded a thorough review of the fusion program by AEC with a report due by early 1966. In May 1965, the AEC formed a committee of scientists chaired by Samuel Allison (MIT) and Ray Herb (U. Wisconsin) to review the fusion program. They submitted their report to AEC in December 1965, and it was forwarded to Congress in January 1966.

The AEC still owed the JCAE its own report, however. In mid-to-late 1965, within the CTR branch and in parallel with the Allison/Herb panel, Ruark undertook to prepare this report by himself, with no staff assistance or help from the fusion research community. After several months, he presented his draft report to Research Division director Paul McDaniel, who found it completely unacceptable [4]. McDaniel removed Ruark as head of the CTR branch and convinced Amasa Bishop, who then was doing experiments on the Model C stellarator at PPPL, to come to AEC on special assignment from Princeton to prepare the report. During January to February 1966 (for a time holed up in a Washington hotel during a snow storm that shut down the city), Bishop, myself, and Richard F. Post, from the Lawrence Livermore Laboratory, prepared the AEC Policy and Action Paper on Controlled Thermonuclear Research [16]. The draft report then went through several months of executive branch review by the AEC's General Advisory Committee (GAC), the Office of Management and Budget (OMB), and the President's Science Advisory Committee (PSAC). The comments (and changes) mostly had to do with disagreements over how large a budget the fusion program needed or deserved. The report was finally transmitted to the JCAE on July 11, 1966 by AEC Chairman, Glenn T. Seaborg. This report gave the fusion program a new lease on life but one more based on promise than on results. A month after submitting the report to the

JCAE, Bishop left Princeton to accept once more the position as head of the AEC CTR branch, a post he had held previously during much of the 1950s.

One of the recommendations of the Policy and Action paper was that the CTR branch should establish a Standing Committee of advisors, supported by a series of ad hoc panels to review various aspects of the program. That was subsequently done. Bishop chaired the Standing Committee, whose other members consisted of the project directors from the major CTR labs plus General Atomics and four "civilian" members from outside the fusion community. Two especially influential civilian members of the Standing Committee were Solomon J. Buschbaum (a vice president of Bell Labs) and Lewis Branscomb (director of the Joint Institute of Laboratory Astrophysics in Boulder, Colorado). At one meeting of the committee, I remember Branscomb expressing the view that fusion would eventually be shown to be feasible, but not economic. At another meeting, I remember Buschbaum commenting that if one approach to fusion could be shown to work, there would be several that would work. This latter comment was in the context that we had several "approaches" to fusion and the JCAE was frequently asking why we could not narrow down on the number of approaches.

One of the early issues facing Bishop and the Standing Committee concerned the future of the privately funded group at General Atomics. The group prided itself that it was privately funded, and they had no desire to become another AEC-funded group. However, in late 1966, after GA theorist William Drummond took a professorship at the University of Texas, the TAERF advised GA that they would terminate funding by the end of April 1967 and fund fusion research at the University of Texas instead. Predictably, the GA management came to AEC to explore the possibility of getting government funding. They were insistent that a decision had to be made quickly, something government is not particularly good at. While the CTR branch was pondering what to do, many of the top personnel left GA and accepted university or government laboratory positions. Bishop consulted with the Standing Committee, which advised us to let the group dissolve and be absorbed at other institutions. As the April deadline neared, the only senior person left at General Atomics who had not decided to go elsewhere was Tihiro Ohkawa. He had a pending offer from the Lawrence Livermore Laboratory that he was reluctant to accept, preferring to stay at GA if he could. The Standing Committee advised Bishop to take the position that if Ohkawa wanted fusion funding from AEC, he would have to move to an existing AEC-funded fusion site. My advice to Bishop was that if Ohkawa was considered worth funding, we should fund him where he was and not force him to move. We provided GA a contract ($500,000 as I recall), and the General Atomics fusion program has prospered to this day under government contract.

The Tokamak

One of the ad hoc panels we established was set up to review the toroidal approaches, of which the stellarator was the most prominent in the USA (though there were several others), whereas the British were pursuing a toroidal pinch configuration and the Soviets were developing their own toroidal concept, called tokamak. Marshall Rosenbluth was chairman of the Panel, and I served as executive secretary. One of the members of that panel was Harold Furth, a top fusion research scientist from Livermore who would later became director of the Princeton lab. In the panel's draft report, Furth inserted a statement that tokamaks were not as promising as stellarators because they required an electric current flowing in the plasma (stellarators do not) and that current flowing in the plasma would generate plasma instabilities causing poor confinement. My boss at the time, Amasa Bishop, insisted the statement be removed from the report, not because he did not agree with it but because he thought it was insulting to our Soviet colleagues. Furth would later go on to be a champion of the tokamak in the US program.

Following the 1958 Second United Nations Conference on the Peaceful Uses of Atomic Energy, where the USA, UK, and USSR declassified their fusion programs, the UN's International Atomic Energy Agency (IAEA) began hosting international fusion conferences every 2–3 years. At the 1965 conference in Culham, UK, the Soviets claimed they were seeing better plasma confinement in their T-3 tokamak than was being seen in toroidal experiments, like the stellarator, in other countries. Most scientists attending the conference in 1965 were skeptical that the data was being properly interpreted. At the 1968 IAEA conference in Novosibirsk, the Soviets reported still better results from their T-3 tokamak. The plasma had better plasma confinement (by a factor of 10) and higher temperature (by a factor of 10) than was being observed in other torioidal configurations, such as the stellarator. Psychologically, it was not easy for US scientists to accept the fact that their favorite toroidal concept (the stellarator) was being surpassed in performance and promise by the Soviet invention (the tokamak), and this was also true of the British physicists. At the end of the conference, Lev Artsimovich, head of the Soviet program, and Bas Pease, head of the British program, agreed to seek approval from their governments for a team of British fusion scientists to go to the Kurchatov Institute in Moscow with diagnostics not available at the time in the USSR and to verify or disprove the Soviet claims. Such international extended site visits between the USSR and Western countries had not yet become a common occurrence, however, so it was by no means certain that the visit could be arranged. It was arranged, however, and the British team, led by Nicol Peacock and including Michael Forrest, Peter Wilcock, and Derek Robinson, confirmed the Soviet results. This started what was later dubbed a "bandwagon effect" rush around the world to convert existing facilities to the tokamak configuration or to build new ones. Confirmation of the Soviet tokamak results by the British has recently been chronicled by one of the British participants [17].

During 1969, the AEC CTR branch received tokamak proposals from several US institutions, including General Atomics, MIT, Oak Ridge, University of Texas, and last (but not least) Princeton, which reluctantly volunteered to convert their sacred Model C stellarator into a tokamak. Bishop approved them all, with the blessing of the Standing Committee. The Standing Committee was influential in getting Princeton to agree to convert its stellarator according to historian Joan Bromberg [4].

During the 1960s, while working in the CTR branch, I had come to realize that almost everyone working in fusion held a Ph.D. degree. Consequently, I enrolled in the Physics Department at the University of Maryland and began studying part-time while working full-time and starting a family (three children born in the 1960s). Eventually, in mid-1968 (just as the Soviet tokamak results were becoming the focus of attention), I left AEC and took a job as a research physicist at the US Naval Research Laboratory (NRL), where I was able to complete my Ph.D. thesis in 1971 under a joint arrangement between NRL and the University of Maryland. My job was to build and operate a laser-produced plasma experiment. My experimental thesis involved demonstrating the production of "collisionless shock waves" using what was then the largest high-energy, high-power laser in the country (30 J, 30 ns pulse). It was the first laser-produced plasma experiment at NRL.

Shortly before I left AEC to go to NRL, a scientist/engineer named Robert L. Hirsch, working at the International Telephone and Telegraph research laboratory in Indiana, sent in a proposal to AEC to support a new concept they had been working on called electrostatic confinement. Hirsch said they were observing fusion reactions (neutrons) in a relatively small device and suggested that they would fly this device to Washington in a small plane and demonstrate it as part of their proposal presentation. Since the device contained tritium (which is mildly radioactive) as well as deuterium as the fusion fuel, I had to prepare an analysis and a memo to AEC building security attesting that no one would be harmed if the tritium in the device were to escape and get into the building's central air conditioning system. This I did and it was accepted (something that would never be possible today). Hirsch brought his device into the AEC building at 1717 H Street, NW, in Washington, wheeled it into a conference room, plugged it into the wall, and turned it on. The neutron counter started clicking, and we all looked into the chamber through a port to see a small spherical glowing region in the center of the device. We joked that, literally, we were staring fusion neutrons in the face (another thing we would likely not do today). We brought some theorists from the fusion community to the meeting, who promptly cast doubt on the concept (somewhat unfairly, I thought). In reviewing many proposals over the years, I have observed that it is almost impossible to get a positive review of a proposal to pursue any idea that is not already being worked on in the government's own fusion program. Be that as it may, we eventually decided not to provide support for the proposal. Bishop and I were very impressed with Hirsch, however, and as I was leaving AEC to go to NRL, Bishop hired Hirsch and he took over my duties and office space in the CTR branch in mid-1968. He went to the IAEA fusion conference in Novosibirsk in 1968, heard the Soviet tokamak results there, came home, and lobbied vigorously for starting a tokamak effort in the USA. In my view, he was as influential (and perhaps even more so) than the Standing Committee in getting Princeton to convert their stellarator into a tokamak.

Earlier, Bishop also had hired another new staff member at my recommendation: Ben Eastlund. Ben was one of several talented young scientists coming out of an applied plasma physics program at Columbia University under the direction of Prof. Robert J. Gross. I had heard Eastlund give an impressive talk at an American Physical Society Division of Plasma Physics meeting. Eastlund would later team up with CTR staff member Bill Gough to write a stimulating paper, The Fusion Torch, showing that, in addition to, or instead of, using fusion for electricity, one might be able to use the fusion energy released to break down waste or other materials into reusable constituents [18].

Inertial Confinement Fusion

During the 1960s, the fusion effort in the USA and other countries was essentially totally devoted to exploring the prospects of achieving fusion using magnetic field configurations. However, the invention of the laser in 1960 got a number of people and institutions wondering whether a sufficiently large, high-energy, high-power laser could be used to "ignite" a small pellet containing fusion fuel—a micro hydrogen bomb with low enough energy yield that it could be contained in a chamber. These ideas within the AEC laboratories mostly occurred to H-bomb designers, and the ensuing effort was highly classified secret by the AEC. As an AEC staff member with a security clearance, however, I had access to these efforts and was frequently consulted by AEC headquarters classification officers for advice on fusion-related issues. One major concern of these officials was the question of whether an "enemy" (by which they meant the Soviets) might put a large laser in a ship, sail it into New York Harbor, and set off a hydrogen bomb using a laser. The CIA also paid me frequent visits to advise me of their fears of "rapid progress" being made in the USSR on "laser fusion" based on "intelligence" reports (the laser was coinvented by N. Basov). The members of the CTR Standing Committee also had clearances (as most fusion scientists did in those days) and requested one of their members, Keith Brueckner from the University of California, to chair an ad hoc panel to look into the prospects for laser fusion. Brueckner's panel concluded the area was promising and recommended the CTR branch fund some studies in this area. We did not, partly because funding was tight but mostly because to pursue approaches that were heavily classified in an era when all of the existing approaches had been declassified would have complicated the open international dialogue and spirit of collaboration that had developed since the declassification of Project Sherwood in 1958. Pursuit of laser fusion (or more broadly "inertial confinement fusion") was left to the AEC Division of Military Application.

Brueckner, himself, had other ideas. He did his own calculations on what it would take to make a successful laser fusion device, resigned from the Standing Committee and, in 1969, joined the firm of KMS, Inc. in Michigan. Brueckner and KMS began to file patent applications (because of secrecy, the AEC labs had not filed for patents). KMS president Kip Seigel wrote to AEC Chairman Glenn Seaborg [4] that KMS could prove the scientific feasibility of laser fusion within 18 months and bring "efficient fusion power" into "availability in the next few years." This launched a several year battle between KMS and the government over classification and intellectual property rights.

My knowledge of and the interest in laser fusion coincided nicely with the opportunity being provided to me by Alan Kolb, superintendent of the Plasma Physics Division at NRL, to set up a laser-produced plasma experiment there using the largest high-energy, high-power laser in the country at the time and to do my Ph.D. thesis using that facility.

The Glory Years: 1970s 4

You have to set doable goals, but you still have to have a very lofty vision.
And you need a well-articulated vision that people can follow.

–Steven P. Jobs
Chairman, Apple Computer, Inc.

John Stamper, Ed McLean, and I built and operated an experiment at NRL using the high-power laser beam that was piped over to our building from the building next door. My thesis task was to see if I could demonstrate the existence of collisionless shock waves. Such shocks had been predicted from theory. My experiment was successful and resulted in two papers published in Physical Review Letters [19, 20]. A surprise bonus came to our group when we observed very large signals on a magnetic probe diagnostic designed by John Stamper. At first, we thought the signal was either spurious or a result of a calibration error. But further investigation convinced us it was real, and, with help on a theoretical interpretation from our theorists, we published a paper in Physical Review Letters on the spontaneous production of megagauss magnetic fields in laser-produced plasmas [21]. For this, we received the NRL Research Publication Award in February 1972. During that period, I also analyzed the possibility of confining a laser-produced plasma in a superconducting resonant cavity by RF electromagnetic fields [22].

The NRL laser was owned and operated by the Optics Division, headed by Walter Sooy, within its laser branch headed by John Emmett. Emmett was a brilliant and ambitious individual who knew everything about lasers and nothing (at that time) about fusion. We had several discussions about laser fusion, and he became intensely interested. I wrote a paper "Laser-Generated Fusion Plasmas" published in Report of NRL Progress, December 1971, describing the relevance of NRL laser plasma research to fusion research [23]. I made General Edward Giller, then head of the AEC's Division of Military Applications, and Richard Balzhiser, then assistant director of the White House Office of Science and Technology (OSTP), aware of the NRL capabilities in high-power lasers and plasma physics related to fusion. Balszhiser sent one of his staff members over to NRL to visit, and later a briefing by Emmett

S.O. Dean, *Search for the Ultimate Energy Source: A History of the U.S. Fusion Energy Program*, Green Energy and Technology, DOI 10.1007/978-1-4614-6037-4_4,

was provided to OSTP on the NRL capabilities. Emmett and I put together a proposal to the NRL director of research, Dr. Alan Berman, for a large NRL effort to develop laser fusion at the laboratory. He was not interested, presumably because he felt such an effort was in the jurisdiction of the AEC and not the Navy. Nevertheless, on Christmas Eve 1971, Balzhiser arranged with the White House Office of Management and Budget to put $5 million, designated for NRL, into the AEC's FY 1973 Presidential Budget request. That launched an NRL–AEC connection in laser fusion that has continued (with some ups and downs) to the present day.

Tokamaks

While I was at NRL (1968–1972), the US fusion program changed dramatically, mostly due to the impact of the Soviet tokamak results but also due to changing political views regarding the importance of energy.

In the spring of 1970, Bishop resigned and took a diplomatic post with the United Nations in Geneva. The AEC brought in Roy Gould, a physics professor from the California Institute of Technology, as head of the CTR Branch. Gould and Hirsch completed the task, begun under Bishop in 1969, to construct several tokamaks in the USA. By May 1970, Princeton was doing experiments in the tokamak they built by converting the Model C stellarator and was duplicating and confirming the Soviet T-3 results. As US scientists were building their first tokamaks, they simultaneously began to conceptualize larger tokamaks that could produce actual fusion conditions.

Also as the 1970s began, the US electric utilities and the Executive Branch of the US government began to review future needs for energy supply. Several electric utilities that had their own internal R&D departments provided some funding to fusion researchers. Ray Huse (R&D manager at Public Service Electric and Gas of New Jersey) and Howard Drew (Texas Utilities Services Company and VP of TAERF) were among the leaders. Ed David, who was the president's science advisor and head of OSTP, instituted a review of national energy R&D [4], and, in May 1971, David sent a request to AEC Chairman Glenn Seaborg asking him to outline two cases for fusion development: a "significantly expanded" case and an "all out program that seeks to develop fusion energy in the shortest feasible time." Richard Nixon was president (1969–1974) at the time and was interested in US "energy independence."

In the summer of 1971, Glenn Seaborg resigned as AEC chairman and was replaced by James Schlesinger. In December 1971, Schlesinger upgraded the CTR Branch to division status, calling it the Division of Controlled Thermonuclear Research (DCTR), reporting to AEC Assistant General Manager Spofford G. English. Having received my Ph.D. in physics in June 1971 (with thesis advisor Hans Griem) at the University of Maryland, and having been in regular contact with my AEC colleagues, I returned to AEC in February 1972 at the invitation of DCTR Director Roy Gould. Shortly thereafter, John Emmett left NRL and joined the Lawrence Livermore Laboratory, where he proceeded to oversee the construction of several ever-larger lasers and assembled an impressive staff of scientists (some of whom came from NRL) to work on laser fusion.

When I left AEC in 1968, the FY 1968 fusion budget was $27 million. When I returned in February 1972, the FY 1972 budget was $33 million. This does not sound like much today, but it represented a 22% increase and allowed for the transition to several new, though small, tokamak facilities. Also, I found the AEC fusion office (in response to the OSTP request mentioned above) immersed in planning (mostly being prepared by Bob Hirsch) for the future and, in particular, trying to define future facilities that would demonstrate the "scientific feasibility" of fusion. This term does not have a rigorous definition but, at the time, was generally accepted to mean operating an experiment that achieved the Lawson parameters in deuterium plasma (without the added expense of handling tritium as fuel). The Lawson parameters define the minimum conditions required for a self-sustaining fusion reaction: temperature at least 5 keV and density-confinement time product approximately 10^{14} ions per cubic centimeter-seconds. Another accepted definition of "scientific feasibility" that evolved was to operate an experiment that would achieve "equivalent scientific breakeven," by which was meant to contain a (non-tritium-containing) plasma that would have produced (if tritium had been used) as much fusion energy as the energy input for heating the plasma. Others, primarily Hirsch at DCTR, wanted to plan for a full (deuterium–tritium) breakeven experiment. John Clarke and Mike Roberts at the Oak Ridge laboratory were encouraged by Hirsch to undertake the conceptual design of such a facility (without much encouragement from Oak Ridge fusion project director Herman Postma). In general, the AEC laboratory fusion program directors did not like the idea of putting radioactive tritium in their experiments.

Planning

When I arrived in DCTR in February 1972, I soon got caught up in an AEC-wide exercise under the AEC Policy and Planning Division which called for all AEC programs to evaluate themselves using a process called "management by objectives" which included the use of "decision tree analysis," a formalism that was popular at the Office of Management and Budget (OMB) at the time. Fortunately, Hirsch had already given the subject much thought and had provided the broad outlines of a fusion development plan [4]. As a part of that exercise, I prepared a chart showing the potential flow of the various elements of the fusion research program and major facilities (current and future), highlighting the timeline and decision points for major decisions. A copy of the poster-sized chart, dated October 1972, still hangs on the wall in my office (Fig. 4.1).

The chart shows a primary competition among three magnetic approaches (tokamak, mirror, and theta pinch), a parallel line for laser fusion (which was not in our purview, but in the AEC weapons program) and supporting lines for basic plasma physics and technology development. For the tokamak, it projected a decision point, based on projected results from the planned Princeton Large Torus (PLT) tokamak

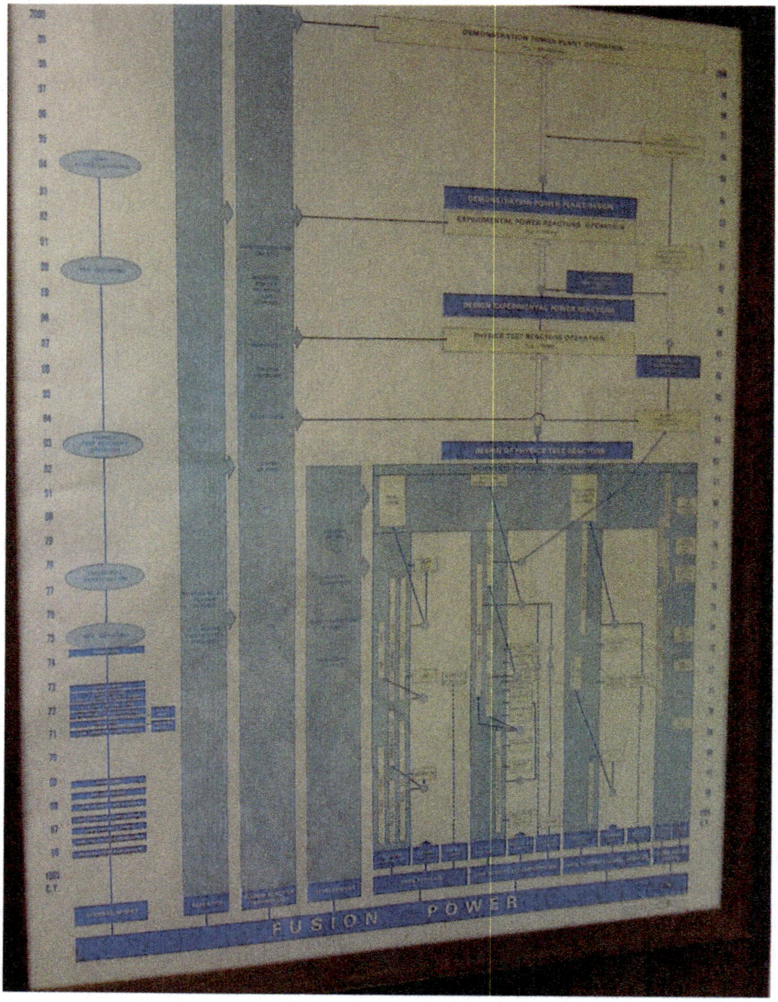

Fig. 4.1 October 1972 fusion power planning chart showing decision points to construct a series of test facilities, culminating with operation of a fusion power demonstration reactor in the year 2000

in 1978, and to construct a Physics Test Reactor that could produce about 10 MW of fusion power, to operate about 1984.

As we shall see, the PLT results from Princeton came in on schedule in 1978, and a decision was made (in 1975 based on anticipated results from the smaller ORMAK tokamak at Oak Ridge and of the eventual PLT results) to construct a major new tokamak called TFTR (Tokamak Fusion Test Reactor) that would be the Physics Test Reactor shown on the chart. The TFTR later would indeed produce 10 MW of fusion power (albeit about 10 years later than we had planned for reasons discussed later).

Other objectives listed were to operate an experimental power reactor around 1990 and a demonstration power plant in the year 2000.

I led a fusion community effort to prepare an R&D plan to reach these objectives which resulted in a detailed (5-volume) program plan in 1976 [24]. This plan was later codified in the Magnetic Fusion Energy Engineering Act of 1980, approved by Congress and signed by President Carter in October 1980 (more on this in the next chapter). Much of the groundwork for doing this planning had been done internally by Bob Hirsch and Roy Gould, in consultation with the CTR Standing Committee, and scientists at the CTR laboratories during the 1969–1972 period.

Management

Although AEC Chairman Schlesinger had elevated the fusion program to division status in December 1971, there was little change in the modus operandi of the fusion program office under Gould. That was about to change. Gould left in the summer of 1972 to return to CalTech. Hirsch was named director of DCTR in August and immediately set out to expand the role of DCTR in managing the fusion program [4]. He established three assistant director positions. I became Assistant Director for (Magnetic) Confinement Systems. He brought in Alvin W. Trivelpiece, a professor from the University of Maryland, to be Assistant Director for Research, and (in 1973) hired Robert W. Bussard, an engineer from Los Alamos Scientific Laboratory, to be Assistant Director for Development and Technology. In August 1971, DCTR had 5 technical staff; 1 year later, there were 16, and by October 1975, there would be 50. In August 1972, the US fusion budget (FY 1973) was $40 million; 5 years later, it would be $332 million.

Both Trivelpiece and Bussard had pedigrees more impressive than mine. Trivelpiece was a professor, had published many scientific papers, and had recently coauthored a textbook, Principles of Plasma Physics, with fellow professor Nicholas A. Krall [25]. Bussard was an engineer from the Los Alamos lab with a long history of contributions and big ideas to the nuclear applications field. He was a driving force and had authored a book on nuclear-powered rockets for space propulsion. I, however, had responsibility for the core of the fusion program. If fusion was to succeed, one or more of the major experimental programs in my purview would have to proceed toward achieving fusion plasma conditions. I had the largest fraction of the fusion budget (though Bussard was laying out future technology budget needs that would become very large indeed if the confinement experiments were successful). I do remember, though, Hirsch commenting to us one day, as the three of us all were arguing for increases in our respective budgets, "Gentlemen, do not measure your manhood by the size of your budget."

Hirsch reconstituted the membership and role of the CTR Standing Committee, renaming it the Fusion Power Coordinating Committee (FPCC). He added his three assistant directors (Bussard, Trivelpiece, and myself) to the committee. This was a message to the laboratory project directors that we were to be considered "managers" and not "staff."

My area (Confinement Systems) was responsible for the evolution of the major (and what were viewed as the most "promising" as ultimate fusion reactors) fusion experiments. The Research (later renamed Applied Plasma Physics) group, under Trivelpiece, was responsible for the fusion theory effort and the smaller, less advanced or more basic experiments at universities and labs. The Development and Technology area was responsible for developing the technologies needed for confinement experiments in both the short term and the long term and for systems studies of future test reactors and power plants. Hirsch used the FPCC as a sounding board for validating our new, more aggressive, pursuit of the fusion goal and to help convince the laboratory fusion program leaders to think more expansively. He guided our efforts, but largely left the detailed decision-making, implementation, and management in the hands of his assistant directors, while he skillfully interacted with and secured support for the program from AEC upper management, other parts of the Executive Branch (OMB, OSTP) and Congress.

Trivelpiece and his team made many important contributions in the 2 years he was there. Three stand out in my mind. He and Bennett Miller, one of his staff, undertook to document the rationale for shutting down two experiments at the Livermore lab: Astron and Levitron. It was the first time that the Washington fusion office had ordered a laboratory to shut down a specific experiment. The laboratory view of Washington was that it was our job to provide them the money and their job to decide what to do with it. To Hirsch, shutting down Astron and Levitron was a demonstration to the laboratories that a change in management style had taken place. Lisa Bromberg [4] chronicles these events. A good history of the Astron saga also has been published [15]. Trivelpiece and Miller also recognized the growing capability of computers and their potential use in simulating complex plasma behavior. They arranged for setting up an advanced computing center at Livermore for the exclusive purpose of working on fusion problems. This center became so popular that it was eventually moved to the Berkeley lab, upgraded, and taken over for use by other research programs at AEC. Fusion then had to ask for its fair share of time on the facility. But that was long after Trivelpiece left to return to the University of Maryland. One of Trivelpiece's hires was Jim Decker who was responsible for the experimental programs. In this capacity, Jim nurtured new ideas that could be studied in smaller facilities and included a class of programs called "alternate concepts" to distinguish them from the major fusion concepts that were being explored in Confinement Systems. In 1977, Jim wrote a comprehensive report on eleven such concepts (DOE/ET-0047). These concepts were considered to have some potential in the long run but were not currently ready for "prime time." Over the years, these were categorized by names such as "alternate concepts," "exploratory concepts," or "innovative confinement concepts." While I supported the need for such efforts, they were not my immediate concern.

Our plans for rapid development of fusion required major advances in technologies such as magnets, materials, and plasma heating. In addition, we needed design studies of future engineering test reactors and demonstration power plants to guide the research. These were the responsibility of Bussard and his team. Among his early hires was S. Locke Bogart, who reported directly to Bussard and hence had an overall perspective on the development. He frequently interacted with my group and made important contributions to the major 1976 program-wide planning document [24]. Other key hires in Bussard's group included Jim Williams (systems), Frank Coffman (systems), Klaus Zwilski (materials), Jack Beal (plasma technologies), and Carl Henning (magnets).

Although tokamaks were the "hot topic" in the early 1970s and the AEC was under constant pressure from Congress to reduce the number of approaches, we did not feel tokamak could yet be the only approach. Two approaches (Astron and Levitron) were eliminated, reducing the number of major approaches to three: open systems (magnetic mirrors), closed systems (tokamaks) and high-density systems (theta pinch and bumpy torus). Management of these three primary approaches was the responsibility of my group. Though bumpy torus, an array of magnetic mirrors arranged in a toroidal configuration, was not a "high-density" system, we managed it in high-density systems for our convenience.

As I began to familiarize myself with my new responsibilities in 1972, I realized that the tokamak area (which had evolved in my absence) needed special attention. So, in March 1973, I formed a panel consisting of myself as chair, with Jim Callen (Oak Ridge) and Harold Furth (Princeton) as co-chairs, and including three other leading scientists: John Clarke (Oak Ridge), Tihiro Ohkawa (General Atomics), and Paul Rutherford (Princeton). Our report, Status and Objectives of Tokamak Systems for Fusion Research [26], became our guide for future decisions and its principles are still valid today.

In our report, we said, "In striving towards this goal (a fusion reactor based on the tokamak), the program would be expected to pass successive milestones in confinement physics, thermonuclear breakeven, and engineering, including:

1. Demonstration of confinement and heating in the plasma physics regime of a fusion reactor.
2. Demonstration of thermonuclear burn sufficient to yield net energy gain.
3. Demonstration of the engineering elements necessary for a reactor."

While developed specifically to guide the tokamak effort, I applied these objectives to all three of the major approaches that were in my area of responsibility. I set out to organize and conduct an experimental program to achieve these milestones. As mentioned previously and discussed later, the first milestone was effectively met in 1978, when the Princeton Large Torus (PLT) heated a well-confined plasma to temperatures above the Lawson "ideal ignition temperature" of 5 keV. Progress toward the second milestone was later achieved in the Tokamak Fusion Test Reactor (TFTR), in the Joint European Torus (JET) and in the Japanese JT-60 tokamak. A complete demonstration of milestone 2 and substantial progress toward milestone 3 is expected in the International Thermonuclear Experimental Reactor (ITER) in the 2020s. More later on these achievements and why it took so long to progress beyond milestone 1.

During 1973 and continuing in 1974, the DCTR managers (Hirsch, Trivelpiece, Bussard, and myself) were in constant dialogue with the fusion program project directors (heads of the fusion programs at the major AEC sites) and with the Fusion Power Coordinating Committee regarding the choice of facility to meet milestone 2 above.

Dr. Dixie Lee Ray, a marine biologist, had become AEC chairman in February 1973. She was a strong fusion supporter; hence, we were optimistic that we could ask for a large new construction project. Congressman Mike McCormack from Washington state, a chemist by background, was becoming an influential supporter in Congress and would play a key role during the decade, as I shall describe later. Attention to fusion was also heightened by the 1973 oil embargo that began in October 1973 and lasted until March 1974, which produced long gas lines in the USA and resulted in greater political awareness of US dependence on foreign sources of energy.

There were two schools of thought on what kind of next step fusion experiment to propose. One, primarily advocated by Princeton Plasma Physics Laboratory and its director, Mel Gottlieb, was that a device should be built capable of achieving "equivalent breakeven," that is, one that achieved the Lawson Conditions of temperature, density, and confinement in deuterium but did not use tritium (and therefore did not produce a breakeven level of actual fusion power). The other, put forth by the Oak Ridge Laboratory, was that the experiment should use tritium and produce actual breakeven levels of fusion power. The latter device would obviously cost a lot more, take more time to construct, and be more complicated to operate (due to the necessity of handling tritium, which is radioactive, and the fact that the larger number of fusion neutrons produced would lead to induced radioactivity in the vessel wall material). The fusion project director at Oak Ridge, Herman Postma, was lukewarm to his own proposal. The Oak Ridge design and proposal were primarily due to the interest and work (encouraged by Hirsch) of two members of Postma's group: John Clarke and Mike Roberts.

It seemed clear that only the tokamak concept was ready for serious consideration. The project directors at the mirror and high-density laboratories (Chester Van Atta and later Ken Fowler at the Livermore lab and Dick Taschek and later Fred Ribe at Los Alamos) understandably felt that their programs would be (financially) threatened if a large expansion of the tokamak effort went forward. Even within the four main tokamak groups (Princeton, Oak Ridge, MIT, and General Atomics), there was concern about the possibility of "winners and losers" depending on where the new facility might be located and managed.

Our preference in DCTR was for the most ambitious step forward: a deuterium–tritium burning fusion plasma that would actually produce as much fusion energy as was needed to "ignite" the reaction. The Oak Ridge device fit our desires nicely. However, Princeton was our "premier" fusion laboratory and home to the most influential fusion scientists. Marshall Rosenbluth, affectionately known in the community as "the Pope of Plasma Physics," had gone to the Institute for Advanced Study at Princeton University when he left General Atomics in the 1960s. We considered his endorsement of the new facility to be essential.

Soon after taking office in the fall of 1972, I began hiring staff. Among my early hires were Bob Scott, Sibley Burnett, Anne Davies, and John McBride. I eventually organized my area (Confinement Systems) into three branches: open systems (mirrors), closed systems (tokamaks) and high-density systems (theta and reversed-field pinches and bumpy torus), each headed by a branch chief.

I assigned to each the responsibility to manage the programs in their respective areas, under my general supervision. By 1975, I had 10 technical staff: Arthur Sleeper (scientific coordinator, who replaced John McBride), Anne Davies (who replaced Sibley Burnett in 1974 as tokamak systems branch chief), Ron Blanken, Dave Ignat, John Willis, Ken Moses (high-density systems branch chief), Abe Kadish, Erol Oktay, Bill Ellis (open systems branch chief), T.V. George, and Milt Johnson. Bob Scott soon left us to join the Electric Power Research Institute (EPRI). The complete DCTR organization at the end of 1975 is shown in Table 4.1. By then, Bussard had left and was replaced by Jim Williams; Trivelpiece also had left and was replaced by Bennett Miller.

As scientists, we were often frustrated with the AEC bureaucracy. There were many "support" divisions that viewed themselves and their rules to be paramount. We, of course, felt just the opposite. The first of these I had a run-in with was personnel. The AEC was exempt from the Civil Service personnel rules that governed the rest of the government. However, it had set up a similar grading system for various jobs and eligibility guidelines of those jobs. When I became an assistant director of DCTR, I found that I was entitled to a more "senior" secretary than the very capable one (Ms. Ruth Ann Watkins) that I currently had. I was told to hire a more experienced person from a list I was provided of "eligible" candidates. I objected to this. I did not see why I could not retain my secretary (at her current grade level and salary) and let her be promoted to a higher level over time. Hirsch did not quite agree with me, but he agreed to go with me to see AEC Assistant General Manager Spofford English. After listening to my arguments, English agreed to seek an exception from the head of personnel and was successful. A similar situation arose later when Sibley Burnett, my closed systems branch chief left and I wanted to put N. Anne Davies into the post. She was a talented young woman Burnett and I had hired into the branch previously and who had been doing an outstanding job. Personnel told me she had not yet served in her staff position long enough to be eligible. Again I stood my ground, and again I won. There were other times I found that the administrative bureaucracy would back down if the technical bureaucracy stood its ground. Unfortunately, I have found that this rarely happens today.

Other areas of "conflict" involved security and classification. Since 1958, the AEC's fusion program was supposedly unclassified and cooperating with fusion programs in other countries. International meetings, visits, and personnel exchanges increased markedly during the 1960s, and this trend was increasing still in the 1970s. For historical reasons, much of the AEC fusion program, especially the larger experiments for which I was responsible, were located in AEC's weapons laboratories: Oak Ridge, Los Alamos, and Livermore. Our facilities were located within fenced-in security areas, and procedures for visitor access were cumbersome and sometimes led to embarrassing denials of access for foreign fusion scientists. I spent many hours trying to convince security people that the security boundaries could and should be relocated so that our facilities could be accessed by foreign visitors. Eventually, we wanted to have some foreign scientists working on our experiments on long-term assignment and had to overcome rules at Oak Ridge and Los Alamos that forbid scientists from certain countries from living in towns near those facilities.

Table 4.1 Division of controlled thermonuclear research, 1975

R. L. Hirsch, Director
G. Burdette, Secretary to the Director
E.E. Kintner, Deputy Director
R. Weller, Secretary to the Deputy Director
M. Katz, Technical Assistant to the Director
R. Bingham, Coordinator for Plans
C. Hess, Senior Scientific Advisor
J.R. Young Assistant Director of Administration
S. O. Dean, Assistant Director for Confinement Systems
A. Sleeeper, Scientific Coordinator
R. A. Watkins, Secretary to the Assistant Director
N. A. Davies, Chief, Tokamak Systems Branch
R. Blanken, Scientist
D. Ignat, Scientist
J. Willis, Scientist
P. J. Showe, Secretary
K. G. Moses, Chief, High Density Systems Branch
A. Kadish, Scientist
E. Oktay, Scientist
L. Worden, Secretary, High Density Systems Branch
W. R. Ellis, Chief, Open Systems Branch
T.V. George, Scientist
M. Johnson, Scientist
S. Gilbert, Secretary
J. M Williams, Assistant Director of Development and Technology
L. Bogart, Scientist
J. Haars, Secretary to the Assistant Director
K. M Zwilsky, Chief, Materials and Radiation Effects Branch
M. Cohen, Scientist
E. Dalder, Scientist
C. Finfgeld, Scientist
E. Dalder, Scientist
T. Reuther, Scientist
E. Ruby, Secretary
F. E. Coffman, Chief, Program Management Branch
R. Kostoff, Scientist
M. Murphy, Scientist
J. Neff, Scientist
B. Twining, Scientist
L. Moser, Secretary
J. W. Beal, Chief, Plasma Engineering Branch
H. Cullingford, Scientist
S. Staten, Scientist
C. Smith, Secretary
C. D. Henning, Chief, Magnetic Systems Branch
D. Beard, Scientist
E. Ziurys, Scientist
L. Harman, Secretary
B. R. Miller, Assistant Director for Applied Plasma Physics
R. Stevens, Secretary to the Assistant Director

R. E. Price, Chief, Fusion Plasma Theory Branch
O. Manley, Scientist
D. Priester, Scientist
W. Sadowski, Scientist
J. Hurd, Secretary
J. F. Decker, Chief, Experimental Plasma Research Branch
W. Dove, Scientist
G. Mischke, Scientist
P. Stone, Scientist
R. Hatcher
G.R Ingram, Chief, Computer Services and Technology Branch
J. Esworthy
J. N. Grace, TFTR Project Officer
D.J. McGoff, Chief, Program Management Branch
J. Turi
R. J. Impara, Chief, Systems Engineering Branch
W. Marton
C. Smedira
L. Price, Chief, Reactor Engineering Branch
A. Dixon, Secretary

Even at our own offices in suburban Washington, we were often unable to receive visitors from some countries.

The emerging area of inertial confinement fusion presented other problems associated with the fact that much of the US work was still classified. Though that work was not in my (or DCTR's) area of responsibility (being funded by AEC's Division of Military Application), we took an active interest in trying to get as much as possible declassified. Others, notably the Soviets, were publishing papers containing data that was still considered classified by AEC. The AEC Classification Division, headed by Charles Marshall, had classification guidelines for data they considered relevant to nuclear weapon design. Each of the laboratories had classification officers who reviewed every paper or conference talk, but still had to have a further review in AEC Headquarters. The laboratory scientists were anxious to publish and speak at open meetings about their work, so there was a constant stream of documents coming through headquarters. Marshall was sympathetic to declassifying as much as reasonable. The Division of Military Applications staff was very conservative, so Marshall would usually come to me for a second opinion. A breakthrough in inertial confinement classification occurred in the early 1970s, resulting in John Nuckoll's (Livermore) publication in Nature magazine [27] of a summary of the work that had previously been carried out on a classified basis. Nevertheless, work on inertial confinement fusion continued to be "born classified," and declassification was a constant struggle throughout the 1970s and into the 1980s. The ICF work at the US weapons laboratories is now largely (but not fully) declassified and a healthy international effort in this area is now flourishing.

In 1973, President Nixon signed the "Nixon-Brezhnev Agreement" on cooperation on the Peaceful Uses of Atomic Energy. The first visit to the USSR under this

Fig. 4.2 Visit to Moscow 1974: Al Trivelpiece, Evgenii Velikhov (head of the Soviet program), and the author (Kurchatov Institute photo)

new agreement was a fusion team we put together to go there in 1974 (Fig. 4.2). We were briefed by US security officials on what to expect: searches of our baggage, hidden recording devices and cameras in our hotel rooms, phone calls in our hotel rooms from prostitutes with the aim of photographing us in compromising positions for the purpose of later blackmail, etc. We were advised not to take pictures from the windows of planes or trains or to take pictures of government buildings, including airports and train stations. Needless to say, we became a bit apprehensive. Actually, soon after arrival in Moscow, one member of our group showed symptoms of a nervous breakdown and had to be taken to the US Embassy and flown home.

On one trip, Al Trivelpiece and I had a flight to Moscow that was different than the others. People in our group were arriving on different flights all day and being picked up and escorted to the hotel by USSR government officials. The flight Al and I were supposed to be on had a problem and we were switched to a different flight. The Soviets were unaware of this; consequently, no one was there to meet us when we arrived that evening. Furthermore, we had not been told what hotel we were supposed to be staying at. Fortunately, Al had the home phone number of someone from the US Embassy. That person did not know what hotel we were supposed to be at, but he had a hunch. The Embassy person picked us up, took us to the hotel, and we found that they were expecting us. Al and I were very hungry by then but found the hotel did not have a restaurant. So we went for a (long) walk, looking for a restaurant. In those days, there were not many restaurants, and those few only took people who had made reservations many months in advance. We were denied entrance into the few we found. It did not help that we did not speak Russian. Finally, we arrived at the large Rossia Hotel and found a restaurant upstairs. It was late by then and the restaurant was closing down. The waiters were clearing tables and I saw they were stacking up bread near to where we were standing. I suggested we just take (steal?) a few pieces, which we did. I wondered afterwards what we would have done if we had been arrested.

During our lab visits in Moscow, Leningrad and Kharkov, we noted the presence of many security staff standing on stairwells making sure none of us went wandering off. We were escorted everywhere by government car, accompanied by at least two government officials, one of whom we assumed to be KGB. We were given tours of cities, many lunches and dinners, so much so that we "complained" that too much time was being spent away from our main purpose, which was to have scientific discussions and see the equipment. It was traditional, for example, to have three-hour lunches with much vodka toasting.

On one visit, Trivelpiece and I were being briefed by Dr. Leonid Rudakov, a Soviet expert in pulsed power electron beams. He was known to be exploring their potential use for inertial confinement fusion, similar to a program in the USA at Sandia National Laboratories. He went to the blackboard and described a technique that was highly classified at the time in the USA. Al and I just sat there and occasionally raised eyebrows at each other. The technique, using a small metal canister called a "hohlraum," was later declassified in the USA and is an integral part of what is now known as "indirect drive."

When I took over responsibility for Confinement Systems, one of the first problems I inherited was that the fabrication of the Alcator tokamak at MIT had run into problems. The design and fabrication of the device had been approved in January 1970 and was under the direction of its principal advocate, Prof. Bruno Coppi. Coppi was an outstanding theoretical physicist who also considered himself an engineer. Unfortunately, when the device was assembled in 1972, it was found that it would not operate properly due to mistakes in its engineering design. MIT was not making progress in correcting the problems. We had a confrontation with MIT officials in mid-1973, the result of which was that Prof. Louis Smullin, head of the Electrical Engineering Department, was given overall responsibility for making the necessary changes. These included putting Ronald Parker in charge of the Alcator program, replacing Coppi. The device was torn apart and reengineered with a new vacuum vessel and other changes. It began operation again in 1974 and was very successful.

A more serious problem arose associated with the new Princeton Large Torus (PLT) tokamak that was under construction at Princeton. After the C stellarator had been converted quickly into a tokamak (called the symmetric tokamak, or ST) in 1970 and used to reproduce the Soviet T-3 results, the AEC authorized the fabrication of a larger tokamak, PLT. A somewhat smaller tokamak, ORMAK, was also being built at Oak Ridge. The T-3 and ST tokamaks used ohmic heating (current induced in the plasma) to raise the temperature to about 1 keV. That was about the limit expected to be possible with ohmic heating; so some auxiliary heating technique would be necessary to raise the plasma temperature into the fusion regime of 5 keV or above. The ORMAK was the first tokamak designed to demonstrate auxiliary heating (by injecting a high-energy neutral beam of particles into the plasma). Oak Ridge had been carrying out an aggressive neutral beam technology development program. A similar, though somewhat smaller, neutral beam technology development program was being carried out at the Lawrence Berkeley Laboratory for use to inject particles into magnetic mirror devices at the Livermore Laboratory. The PLT was scheduled to operate in 1974. When I took office in 1973 and reviewed the Princeton budget proposals, it was obvious to me that Princeton had not budgeted for the neutral beams nor did they indicate how they intended to procure them. The beams necessary to heat PLT to 5 keV were well beyond the state of the art. When I asked Princeton Project Director Mel Gottlieb about this, he replied that the beams would be developed and built within the Princeton lab and added to the completed facility as they became available. There was no rush or problem from his point of view. I should just leave the matter in his capable hands. This was completely unacceptable to me and constituted one of the early tests of the new management philosophy instituted by Hirsch. I determined that Princeton did not have the in-house capability to develop and manufacture the beams necessary on the schedule we wanted nor at the size we needed to achieve the 5 keV temperature needed to demonstrate that the Lawson ideal ignition temperature had been met or surpassed. I directed Princeton to talk to the Oak Ridge and Berkeley labs, get proposals from them, and select one of them to develop and deliver the needed beams as rapidly as possible. We would then provide the necessary funding. Princeton eventually selected Oak Ridge.

The atmosphere in the mid-1970s was such that we felt we could propose and likely get funding for a large increase in the size of the AEC fusion effort so as to explore the physics broadly and the technology development aggressively. One consequence of this new perceived freedom for the Confinement Systems program was that Princeton and General Atomics each proposed to build additional new tokamaks that had noncircular plasma shapes (all previous tokamaks had circular shapes). Princeton proposed a D shape; General Atomics proposed a kidney shape. An evolving, more-focused fusion theory effort had predicted that these shapes would not only provide better confinement in a given size of device but also be more compatible with power plant operation from the engineering point of view of impurity control and fueling the plasma. Except for the different shapes, the two devices were similar in size, capability, and cost. A review favored the Princeton proposal if only one device were to be built. I decided to build both. My decision was heavily influenced by nontechnical considerations. I believed that in order to keep to our new schedule and development-oriented fusion plan, we needed to encourage and develop fusion capability in industry. The General Atomics Company was, at the time, marketing gas-cooled fission nuclear power plants. I reasoned that if they were successful and had a good fusion capability, they would be in a position to more rapidly move fusion into its commercialization phase. Also, the "culture" at GA was different from that at Princeton, where the focus was much more on understanding plasma physics than on how to proceed rapidly toward a fusion power plant. One of the main criticisms of supporting a new facility (called Doublet III at the time) was that they did not have sufficient plasma physics staff to carry out the project. The fusion group at General Atomics under Tihiro Ohkawa, which I had a hand in saving from extinction in the mid-1960s, was still very small in the early 1970s compared to the other AEC fusion lab programs and would have to be expanded greatly to take on what was considered to be a major tokamak program. Recalling that at one time in the early 1960s GA had employed some of the best plasma physics talent in the USA, my view was that they would be able to attract whatever talent was needed. In addition, Ohkawa was a very inventive scientist, always thinking of new ways to get to fusion. At one point, he had "invented" laser fusion independently. Since he was Japanese and had no security clearance, the reward for his efforts was a visit from the FBI who said his calculations were considered classified by the AEC. They took his laser fusion papers and advised him he was not to work further on the concept.

General Atomics rose to the occasion and has since done outstanding scientific work, although the company failed in its attempt to commercialize gas-cooled fission reactors and so never took up the fusion commercialization challenge. Though they also quickly abandoned the kidney-shaped plasma in favor of the D shape (Fig. 4.3), there was plenty of physics to be done and the two groups (Princeton and General Atomics) rarely conflicted in the topics they studied. Ohkawa continued to invent and, at one point, secured funding from Phillips Petroleum to develop a reversed-field pinch concept called OHTE. Unfortunately, the stock market tumbled, Phillips Petroleum was sold, and that project died.

Another issue I confronted during the early days of taking up my Confinement Systems (CS) responsibilities was securing theory support for the design of CS experiments and the interpretation of experimental data. Under the DCTR organization chart, major experiments were in my jurisdiction and fusion theory was under the jurisdiction of the research program under Al Trivelpiece (later succeeded by Bennett Miller). Historically, the fusion theorists were a breed apart and insisted they be allowed to pursue their own topics without interference from anybody. Also, within the DCTR, they were funded by a budget that was separate from the CS budget and based on work statements that had been approved by the Research group. So, both at the labs and within DCTR, the theorists had to be "asked and cajoled" to spend time working on experimental issues (which often conflicted with the time they thought was better spent on theoretical topics of interest to them). Often, but not always, these conflicts were resolved at the local level and theorists did provide some (but not enough) support to the experimental effort. I therefore authorized the CS experimental programs to use CS funds for whatever additional theory support they felt they needed to carry out a successful experimental program. I established within my group the post of scientific coordinator and hired a theorist to fill the position, with the charge to look at the theory aspects of all our experiments (not just tokamaks) and to ensure that theory was properly taken into account and provided to our experimental efforts. This post was first filled by John McBride and later by Arthur Sleeper.

Fig. 4.3 Bob Hirsch, Anne Davies, and the author at the General Atomics Doublet Tokamak construction site (General Atomics)

Several program meetings were held during 1973 to discuss the next step tokamak facility beyond PLT. These meetings generally involved Hirsch; his three assistant directors (Trivelpiece, Bussard, and myself); the project directors of the main fusion programs: Mel Gottlieb (Princeton), Herman Postma (Oak Ridge), Dick Taschek (Los Alamos), and Chester Van Atta (Livermore); members of the CTR Standing Committee, of which the most influential was Solomon J. Buschbaum (Bell Labs); and various staff from DCTR and the labs as deemed necessary. The laboratory project directors were at first still coming to grips with the idea that decisions (on what directions their research would take) they previously made were now, in some cases, going to be made in Washington. At one of the first of these meetings, Hirsch called an executive session (managers only; staff excluded). As we were walking into this meeting, Herman Postma turned to me and reminded me that staff was not invited to this meeting, to which I replied, "Herman, I have news for you, I am not staff."

Two of these general management meetings in 1973 were especially significant. One was in the nature of a "retreat" held in Key Biscayne, Florida, in July to try to reach consensus on whether the next device would or would not involve using tritium to produce actual fusion conditions or whether it would be a hydrogen experiment to produce "equivalent breakeven" conditions in the plasma. This was to be a small "closed" meeting to thrash out differences of opinion among the major players. We had one "gate-crasher" however. The staff director for the Congressional Joint

Committee on Atomic Energy (JCAE), Ed Bauser, got wind of our "secret" meeting. Wondering what we were up to and whether we were plotting to spring on them a major new fusion initiative with large funding implications, he simply arrived. Hirsch politely and patiently explained to him what we were doing and, seemingly satisfied, he eventually went back to Washington. At the beginning of the meeting, none of the project directors (even Postma from Oak Ridge, which presumably had the most to gain) favored a tritium-burning experiment. As the discussions continued (and with the endorsement of CTR Standing Committee member Buschbaum), a compromise was reached that endorsed both new hydrogen experiments "which are more simply and economically examined in non-burning plasma" and also a potential deuterium–tritium (DT) burning experiment: "The program should seriously plan for DT burning experiments at a time earlier than previously anticipated." Both sides thought they had won the argument [4].

Tokamak Fusion Test Reactor

During the summer and fall of 1973, as I continued to work with key scientists at Oak Ridge, Princeton, and General Atomics preparing our report, "Status and Objectives of Tokamak Systems for Fusion Research," Hirsch was preparing the AEC Commissioners and the JCAE staff for the possibility of launching a DT burning breakeven construction project, based on our expectations for neutral beam heating experimental results first from ORMAK and then from PLT.

The matter was brought to a head at a December 1973 meeting in Washington. Clarke and Roberts from Oak Ridge presented their design for a full DT breakeven device. The project directors were appalled at its size and cost (estimated at upward of $300 million, while we had budgeted only $100 million for the device). Though Princeton Director Gottlieb was still arguing for a hydrogen experiment (he was adamantly opposed to having radioactive tritium at the Princeton site), his key experimentalist, Harold Furth (who was co-chairing the Status and Objectives study), stepped up to the blackboard and said, "Well, if all you want is DT breakeven, here is how to do it." He sketched out a device (called a "wet wood burner") that achieved DT breakeven by injecting a beam of high-energy deuterons into a low-temperature tritium plasma. Practically no one thought that this concept could extrapolate to a fusion power plant, but, for a breakeven experiment, it might well serve the purpose. The idea caught on quickly and satisfied just about everyone (except Clarke and Roberts). Princeton was told to do further design and come up with a detailed proposal that could be put up for a decision (which we would make in Washington) against the Oak Ridge proposal. Regardless, of which design was chosen, we would make a separate decision on where to site the experiment.

In early 1974, we put the final touches on our Status and Objectives report and published it as AEC report WASH-1295 [24]. It summarized all the known physics and technology issues associated with tokamaks, as best we knew them at the time. In early 1974 also, Herman Postma was named director of Oak Ridge National Laboratory and John Clarke succeeded Postma as project director of the fusion program there. Clarke and Roberts continued the design of their full DT fusion experiment. Princeton, meanwhile, launched a conceptual design for their wet wood burner, which they estimated would cost about $80 million. During the spring of 1974, I had several meetings with Mel Gottlieb at Princeton. He was still hopeful that a new, larger, hydrogen "equivalent breakeven" device would be built and located at Princeton. He did not want a tritium experiment at his lab. Eventually, I just told him, "Look, Mel, we are going to try to get funding for a DT device and there is not likely to be money also for a hydrogen device in the near future. If you don't offer to host the DT device at Princeton, you will likely be left behind as fusion is developed." I am sure he had discussions with many others. However it happened, Gottlieb eventually changed his mind and Princeton offered to host the wet wood burner if it was chosen but not the larger Oak Ridge device if it was chosen (it was obvious to all that the fusion power from the Oak Ridge device would be too large to be licensed at the Princeton site).

In July 1974, DCTR hosted a large meeting in the AEC auditorium to hear presentations from Oak Ridge, Princeton, and other interested scientists and institutions. Our main concern was the presentations from the theory scientists, especially Marshall Rosenbluth. The theorists had a habit of predicting instabilities that would shorten the plasma confinement in any fusion device. To our amazement, none of the theorists made any predictions that either of the devices proposed would fail to perform. The meeting was attended by members of the newly constituted Fusion Power Coordinating Committee (FPPC), which Hirsch had constituted to replace the CTR Standing Committee, though its membership had much in common. In consultation with the FPCC and taking into account our budget projections and desire to have a device operating as quickly as possible for the DT demonstration, we opted for the Princeton design. The decision of where to put it remained open. I knew that Bob Hirsch wanted to site the device at Oak Ridge, even though it was Princeton's design. If we sited it at Oak Ridge, it would result in a buildup of capability there that was more consistent with our desire to then move rapidly ahead with fusion engineering development at the same site, since Oak Ridge was a major nuclear engineering laboratory whereas Princeton was a plasma physics laboratory with no obvious future role in fusion engineering development. My view was that what was most important was that the experiment be successful in achieving its purpose and that Princeton scientists were best qualified to accomplish that. I felt that a fusion engineering test reactor could still be sited at Oak Ridge later, especially if we continued to build up the fusion technology development program there. I was also influenced in my view by the fact that the design we had chosen was Princeton's design and therefore, I believed, they would be more likely to "put their heart into it" to make it a success. In any case, after our internal discussions, Hirsch decided to locate the device, to be called the Tokamak Fusion Test Reactor (TFTR) at Princeton. We were successful in putting a request for TFTR construction in the FY 1976 AEC budget, developed during the summer of 1974, that was sent to Congress in January 1975. It was subsequently approved. The official goal of TFTR, as established in March 1976, was to produce 1–10 MJ of fusion energy.

The political situation in Washington was changing rapidly in 1974. Richard Nixon resigned in the wake of the Watergate scandal and Gerald Ford became president. Dixie Lee Ray remained AEC chairman but the antinuclear movement was growing in influence and Congress decided that AEC should not be both nuclear advocate and nuclear regulator, so they removed regulatory authority from AEC by setting up a new agency, the Nuclear Regulatory Commission. They then broadened the scope of AEC to include all of energy and not just atomic energy by abolishing the AEC and replacing it with a new agency, the Energy Research and Development Administration (ERDA). All the existing office space, staff, laboratories, etc. (including nuclear weapons programs) transferred to ERDA officially on January 19, 1975.

Though tokamaks had assumed prominence by 1972, I still had two other competing approaches (open systems and high-density systems) in the CS portfolio, and they were not without issues of their own. It was my job to give these other two fusion approaches a fair shot at showing their potential to meet milestone 1 in a timely way, that is, not too far behind the tokamak, which was pulling ahead rapidly.

High-Density Systems: Pinches

Both the open systems (magnetic mirror) and high-density systems (pinch and bumpy torus) approaches to fusion were well established. Mirrors and pinches were among the first fusion concepts proposed in the 1950s and 1960s. Open systems work was carried out primarily at the Lawrence Livermore and Oak Ridge laboratories; high-density systems were primarily carried out at the Los Alamos laboratory (pinches) and Oak Ridge (bumpy torus). Having worked in the AEC fusion program office during the 1960s, I was well familiar both areas. Tokamaks, on the other hand, had evolved into prominence, while I was away at NRL. By the time I took over CS in 1972, Oak Ridge had essentially abandoned mirror research in favor of tokamaks; so the remaining large-scale mirror effort was at Livermore.

Los Alamos had two parallel pinch programs: the theta pinch and the toroidal Z-pinch. Under the new system of Ad Hoc Panels set up by Bishop after the 1966 AEC Policy and Action Paper, the Los Alamos program was the first to be evaluated. The panel, chaired by University of Maryland professor Hans Griem (a fusion pioneer since the 1950s, who also became my Ph.D. thesis advisor), advised that Los Alamos build a larger, longer, linear theta pinch than the one they had at the time (Scylla) and plan on building a toriodal one (Scyllac). Los Alamos had already shown that high-temperature, high-density plasmas (compared to tokamaks and mirrors) could be formed by the theta pinch method. The physics problem was that, in linear geometry, the plasma was rapidly lost out the ends and, in toroidal geometry, would be expected to go unstable. From the power plant perspective, the pulsed nature of the approach did not look very attractive. On the other hand, these high-density pulsed plasmas did not seem to be plagued by the myriad of "micro-instabilities" that were being observed in the low-density mirror and tokamak plasmas. The Los Alamos philosophy then was to get a lot of fusion energy rapidly (before instability could quench the reactions) and then pulse repeatedly.

When I returned to AEC in 1972, I found that a 5 m linear machine (Scylla IV) with mirrors on the ends was in operation and giving good data. The toroidal Scyllac had not yet been built. It had been decided to propose controlling the predicted unstable motion of the toroidal plasma using "feedback stabilization" coils in Scyllac. A 120° curved sector had been constructed to test the feedback technique. The sector had started operation in 1971 but without the feedback system. The feedback hardware was still being tested separately. Under what they (rightly) perceived to be a threat from tokamak progress, Los Alamos decided, just as I was taking office, to bypass the sector tests and to proceed immediately to construct the Scyllac, based on theoretical calculations of what was needed for feedback performance and on separate engineering tests of feedback hardware. The Scyllac began operation in April 1974. By October, it was clear that the feedback system, as designed and built, could not respond fast enough to stabilize the plasma.

As indicated previously, the AEC was abolished at the end of 1974 and replaced by ERDA. During 1975, we continued to consider what to do about the Scyllac situation. The feedback could be redesigned and rebuilt but would cost money and result in further delay in the "race" with tokamaks and mirrors. This I would accept if others felt the toroidal theta pinch had reasonable power plant prospects. Jim Williams had replaced Bob Bussard as Assistant DCTR Director for Development and Technology (which was responsible for power plant studies). He felt that "highly pulsed reactors" would not be acceptable to the electric utilities [4]. Hirsch was also of that general persuasion. Within Los Alamos, the Z-pinch program scientists (who had a smaller device) thought that the demise of Scyllac would open the door to their concept. Even within the Scyllac group, some preferred to go back to the linear Scylla IV and work on "end-stoppering" techniques. Fred Ribe, the theta pinch program director, continued to believe in Scyllac.

During 1975 and 1976, Scyllac was operated in a "sector configuration" to study the physics of the interaction of the plasma with the feedback system that was available. We gave additional support to both the linear theta pinch experiments and the Z-pinch program. A new linear theta pinch, called the staged theta pinch, was built under the direction of Keith Thomassen (who would later head the Livermore mirror program), and the Scylla IV was upgraded and renamed Scylla IV-P. Meanwhile, the financial needs of the tokamak and mirror programs, and the relative progress and promise (in my mind) of those approaches compared to the pinch programs, led me to the conclusion that it was time to narrow the number of major fusion approaches in the Confinement Systems area from three to two. The issue came to a head at a Fusion Power Coordinating Committee meeting in December 1976. Bob Hirsch had been promoted within ERDA in early 1976 to be Assistant Administrator for Energy Research; his deputy, Ed Kintner, had replaced him as Director of DCTR. I reported to Kintner, and Kintner reported to Hirsch. I said that I wanted to close out the Scyllac program and also did not consider the Z-pinch program to be sufficiently advanced to remain within the Confinement Systems program either. Kintner did not want to shut down the Los Alamos experimental fusion program completely. We agreed to stop the Scyllac effort and to transfer the Z-pinch to the research program assistant director within DCTR, who by then was Ron Davidson as successor to Bennett Miller. Before finalizing the agreement, Kintner asked me to promise him that I would not go to Bob Hirsch and try to get him to agree to eliminate the Los Alamos fusion effort altogether. I assured him I would not. The thought of "going around" Kintner never occurred to me, though I suspect Hirsch might have been inclined to agree with me.

Bumpy Torus

The bumpy torus had been under investigation at Oak Ridge since the mid-1960s. On one of my visits to Oak Ridge in the 1960s, the laboratory director, Alvin Weinberg, decided to take me to lunch. As we were driving in his car, it was clear that he was very excited to tell me something. That something turned out to be that he considered recent results from an experiment called IMP likely held the key to fusion. The single-cell IMP continued to look interesting throughout the 1960s, even though Weinberg's belief that fusion neutrons were being observed in IMP turned out not to be true. Consequently, in November 1971, Oak Ridge submitted a proposal to build a toroidal version (called Elmo Bumpy Torus or EBT) consisting of 24 linked mirror segments (Fig. 4.4). It was approved and began operation in 1974. By late 1975, the group was claiming density-confinement time products comparable to what was being achieved in the mirror 2XIIB, although at lower temperatures and densities. Diagnostics on the plasma were minimal at first. The next couple of years were spent improving the diagnostics and studying the plasma in some detail. By 1978, Oak Ridge was proposing to build a major new facility (called EBT-P). It looked like we might have a third horse in the race after all.

I saw the EBT-P as the opportunity I had been watching for to bring another industry (other than General Atomics) into the fusion program. I advised Oak Ridge that I wished the EBT-P to be put out for bid for industry to build and operate. I wanted an industry to take "ownership" of the concept and to advocate its development as a fusion reactor. They would, of course, rely on Oak Ridge for advice as needed but would likely actually hire some or all of the current EBT staff. Oak Ridge management was appalled at the idea, saying that I had no right to take away "their project" from them and "give it to somebody else." I pointed out to them that all of "their projects" were actually "owned by the government" and that they operated them at our pleasure. I said that I would be happy for them to oversee the bidding and selection process for the government or we would put the project out for bid from headquarters. Oak Ridge agreed to the former and selected McDonnell Douglas. McDonnell Douglas agreed to build a facility in Oak Ridge at their expense to house the project. When I left government in mid-1979, EBT-P procurement was under way. It would soon suffer a sad fate, however (Chap. 5).

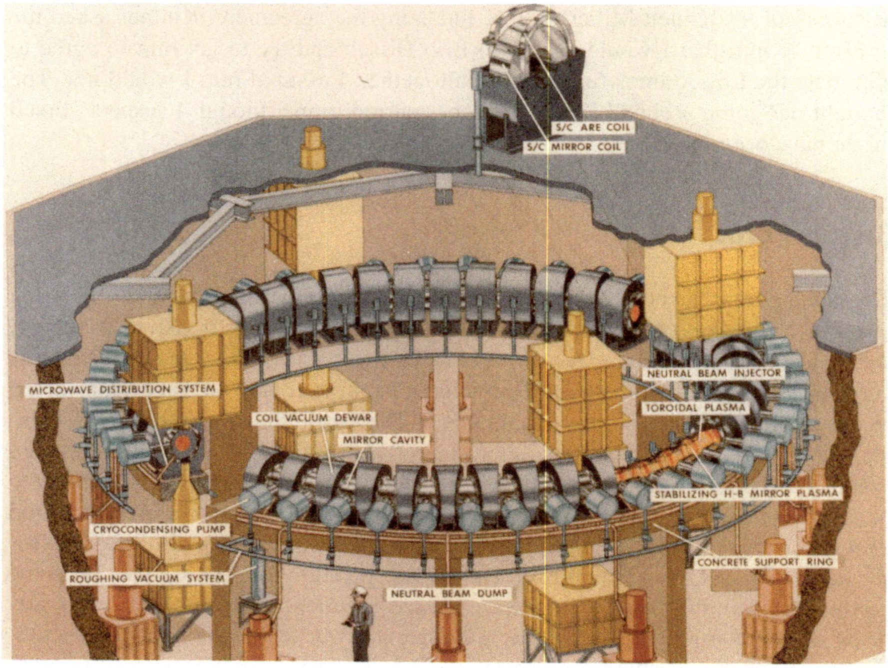

Fig. 4.4 Artist conceptual drawing of Elmo Bumpy Torus (EBT). The magnets surrounding the chamber were energized in such a way to produce a series of linked magnetic mirrors, which distinguished the concept from other toroidal concepts like the tokamak, stellarator, and toroidal pinch (Oak Ridge National Laboratory)

Open Systems: Magnetic Mirrors

In 1972, Livermore had two mirror experiments in operation: Baseball II and 2XII. The first involved injecting and trapping neutral beams of hydrogen into a vacuum tank surrounded by superconducting magnets in the shape of the seam on a baseball. The second involved injecting modest density plasma into a vacuum tank surrounded by normal conducting magnet coils and then compressing the plasma, further raising its density and temperature. The Baseball II, which was favored by Livermore management under Ken Fowler (who had succeeded Van Atta), had for several years unsuccessfully tried to raise the density of the trapped plasma. The theorists loved this experiment, though, because it kept them endlessly engaged in trying to postulate and confirm various theories about what was limiting the buildup. The 2XII plasmas, on the other hand, were prosaic to the theorists. The 2XII plasma behaved largely in a very understandable fashion. The Livermore fusion program management also believed that the Baseball concept, if it could be made to work, could be more readily extrapolated to a fusion reactor. The problem with that logic, however, was that mirror reactor systems studies indicated that, at best, the resulting mirror "reactors" might barely be able to produce net energy, due to plasma loss rates from the mirror regions. I was determined to reverse the priorities at Livermore between the Baseball effort and the 2XII experimental efforts.

In December 1972, I convened a review panel at Livermore to discuss setting near-term goals in the light of the new program strategy being developed in DCTR under Hirsch and in light of the rapid evolution under way toward tokamaks in the world fusion program. I told them they would need to produce plasmas with parameters more closely aligned with those being currently and projected to be achieved by tokamaks. I said I did not see how they could do this via the Baseball approach. At first, Livermore balked at any suggestion that Washington fusion personnel had any business telling the laboratory how to run their program. However, Hirsch was simultaneously in the process of telling the laboratory to shut down two other Livermore fusion experiments, Astron and Levitron, and they were losing the argument. I had useful discussions with lab director John Foster about all these matters and found him to be most understanding of the situation.

Livermore agreed to enhance the 2XII team and upgrade the facility, under the direction of 2XII team leader Fred Coensgen, with the ambitious goal of adding 12 neutral beams to the facility (now to be called 2XIIB). Baseball was to be continued, but at a reduced level. A target date of late 1975 was set for achieving plasma parameters in 2XIIB more in line with those achieved in tokamaks. The hoped-for results came in on schedule in 1975.

The 1976 Magnetic Fusion Program Plan

In early 1975, Hirsch told me he was planning to hire a deputy and asked if I was interested. I said no, I thought my present position gave me more authority as a line manager than to be a deputy. What I did not consider was the possibility that Hirsch might move up in the new ERDA organization, in which case his deputy might be in line to take over as head of the US fusion program. Soon Hirsch had picked the man he wanted for deputy: Ed Kintner, a nuclear engineer who had been a key aide to Admiral Rickover and who was now working as a deputy to Milton Shaw, who headed the AEC's fission breeder reactor program. Before finalizing an offer to Kintner, however, Hirsch had him "interview" with his current division directors, including myself. Kintner, who was 15 years older than I (I was 39 at the time), later told me that he thought it was a bit humiliating for him to be put through the interview process with people he knew were less experienced than he. He, however, knew nothing about plasma physics and fusion and did not hold a Ph.D. degree. He did have an impressive career resume and a strong, yet engaging, personality. Both Hirsch and I were admirers of what Admiral Rickover had accomplished in the 1950s nuclear submarine program. Rickover first asked the AEC laboratories how he should go about designing and building a nuclear reactor to go into a submarine so that the submarine could stay underwater almost indefinitely. The AEC labs laid out a research, development and testing program that would take decades. Rickover wanted the reactors at sea within a few years, so he went to General Electric and

Westinghouse and then launched a successful development effort on his few-year schedule. As a side benefit, his basic reactor designs became the beginnings of the deployment of civilian nuclear power plants. It seemed to me that Ed Kintner had a lot to offer to fusion in view of how quickly we wanted fusion to evolve as an engineering development effort and that his experience might coincide nicely with my own view of how important it was to involve industry in that effort. Ed became Hirsch's deputy in March 1975.

During most of 1975, we continued to implement the plans that had been laid in place for the tokamak program as discussed above. Hirsch continued to focus on getting fusion understood and accepted by "higher up" officials in ERDA, OMB, and Congress. Kintner began to familiarize himself with fusion, making visits to the fusion laboratories and developing relationships with the project directors and other fusion scientists via visits and participating in meetings. Management of the subprograms (Confinement Systems, Research, and Development and Technology) remained largely as before, with Kintner often signing formal approval documents for budget allocations, etc. when Hirsch was absent.

Hirsch wanted to prepare a new fusion planning document that would hopefully become incorporated into ERDA policy planning. He started this activity in December 1975. He noted that because "need, intent, and funding are finally decided by others, the fusion program requires a number of plans by which the program can be conducted." These he gave the name "logics." He then asked me to lead an effort that would involve the rest of the fusion community to complete the task. This I did, resulting in a five-volume comprehensive planning document [24]. The resulting plans (called logics) are shown in summary in Fig. 4.5. The planning document provided detailed budget needs for sub-elements. After the plan was completed and about to be published, the Office of Management and Budget objected to including budgets in the document. In my view, the plans made no sense without the budgets. I presented a briefing to ERDA Administrator Bob Seamans in July 1976, and we asked him to ignore or overrule the OMB objections. This he did and the plan was published. In my experience, this was the first and only time the OMB failed to get an agency to conform to their wishes regarding the public distribution of plans with budget needs included. OMB felt such documents could be used to put pressure on them to provide budgets for approved programs. My view was and is that without the required budgets to implement plans, program managers cannot be expected to deliver on their projected schedules.

By the time the new program plan was completed in July 1976, Bob Hirsch had been named Assistant Administrator of ERDA (in March 1976), overseeing a broad range of ERDA's research portfolio, including fusion and solar, and Kintner had been named to succeed Hirsch as Director of the US fusion program.

During ERDA's short existence (1975–1978), the DCTR was first renamed the Division of Magnetic Fusion Energy (DMFE) and then "elevated" to an office called the Office of Fusion Energy (OFE). Each of the program offices were also "elevated" to division status. I became Director, Confinement Systems Division. The other divisions were Applied Plasma Physics (headed by Ron Davidson and later by Jim Decker), and Development and Technology (headed by Frank Coffman, who

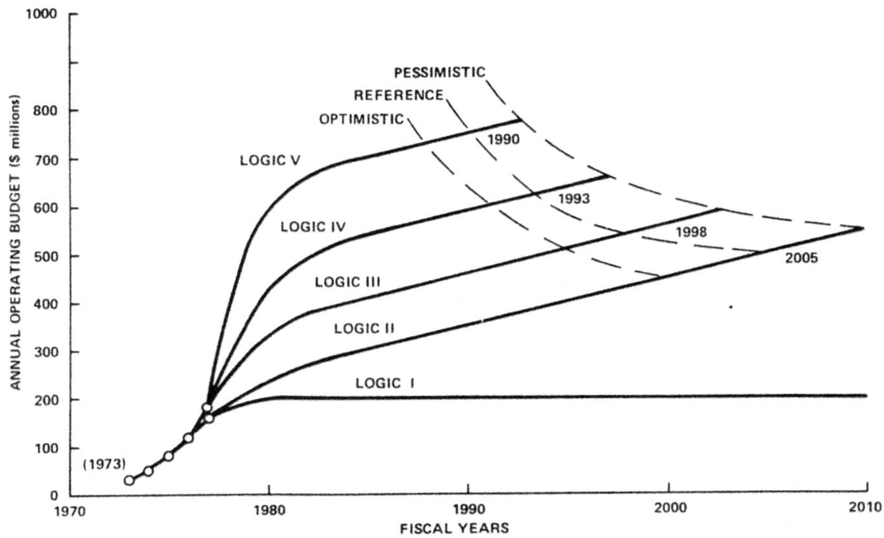

Fig. 4.5 1976 fusion planning logics. The "logics" ranged from I (level of effort research) to V (maximum effective effort). Logic III (aggressive) was chosen as our "reference" case, with a 1998 target date for operating a demonstration power plant (*Fusion Power by Magnetic Confinement: Program Plan,* US Energy Research and Development Administration report ERDA-76/110, July 1976. Posted at http://fire.pppl.gov/us_fusion_plan_1976.pdf. Also published in J. Fusion Energ. **17**(4) (1998)

succeeded Jim Williams in 1978). Kintner hired John Clarke from Oak Ridge to be his deputy. Shortly after the formation of ERDA, he also hired Ron Davidson from the University of Maryland to head the Applied Plasma Physics Division.

It soon became clear that Kintner had much more inclination to deal directly with the project directors at the labs than to work through his division directors like myself. He viewed us more as his "staff" than as "line managers." He also felt he needed to make it clear to the lab directors that he had taken charge. While we continued to evaluate and manage our respective program areas as before, Kintner involved himself more in our subprograms than did Hirsch. He was in almost daily contact with the project directors at the labs, discussing their views and giving them to understand that they only needed to heed orders from him and not from us. He would sometimes authorize them to take actions without our approval which, under Hirsch, they understood they needed. Perhaps he learned this management style from Rickover and/or Shaw, of whom it is said they reserved all decision-making to themselves.

Although my Confinement Systems program was the ultimate "customer" for the TFTR construction project and would eventually be responsible for its operation, Hirsch had decided he should set up a separate "Technical Projects Office" to oversee the construction phase. He brought in Nelson Grace to head up that group. One thing we were all in agreement on was that Princeton would not be allowed to manage the construction "in-house" and to fabricate (as was their past tendency) many of the parts in house and to staff up an engineering department to do TFTR-detailed design and eventual assembly. We advised Princeton that they would be required to bring in an industrial prime contractor that would work under contract to them. Princeton fought us on that idea for a time, lost, and eventually held a competition that resulted in selection of the team of Grumman Aerospace and Ebasco Engineering Services.

In general, I interacted well with Nelson Grace during construction and tried to keep my nose out of his business. I did keep an eye out for anything that might compromise TFTR eventual performance. The only major issue that came up during construction had to do with cost escalation. As with many large construction projects, the original cost was soon seen to be too low. At first, the joke (and it was only partly a joke) was that as the cost crept up, Nelson would start removing chairs from the planned auditorium in the TFTR building, followed by elimination of office space for the scientists. However, at one point, Nelson decided (without consulting me) that he would reduce the number of neutral beams planned for heating the plasma from 4 to 2. When I heard this, I "hit the roof" since this would clearly mean its fusion performance would be compromised. Nelson backed down and the full complement of beams was back in the schedule. The eventual cost, though, did come in at around $300 million instead of $100 million. Could we have had the larger, more ambitious Oak Ridge device for that price? Very doubtful!

US Department of Energy

In the fall of 1976, Jimmy Carter was elected president of the United States. He was very focused on energy and decided to transform the mostly research-oriented ERDA into a cabinet-level agency called the Department of Energy (DOE) that would also have extensive nonnuclear regulatory authority. As is customary when the presidency changes hands from one party (in this case the Republicans) to the other (in this case the Democrats), the ERDA presidential appointees (of which Hirsch was one) would likely be replaced.

As the transition from ERDA to DOE was under way in early 1977, Energy Secretary-designate James Schlesinger met with Hirsch. Tom Heppenheimer, in his book ([36], p.194), says Schlesinger opened the meeting with the quip "Seems like the fusion program has gotten out of hand." Hirsch reportedly replied, "That's your fault," by which he meant that Schlesinger was in charge of the AEC when Hirsch was appointed to head the fusion program so the responsibility was Schlesinger's. According to Hirsch (private communication), "both had a good laugh." Schlesinger was willing to offer Hirsch a job in the new DOE but not as a presidential appointee. Hirsch already was thinking of leaving government and had an offer in hand from Exxon. Hirsch left government in March 1977 and took a senior post in the Science and Technology Department at Exxon headquarters in New York.

Joan Bromberg, in her book ([4], p. 236), states, "Schlesinger was also alarmed by the rapidity with which the CTR budget had mushroomed since he had left the

AEC in 1973. He wondered whether it would not be wise for him to trim $100–200 million from the FY 1979 (CTR) budget for distribution to some of the DOE programs to which he gave higher priorities." She references "Author's interviews with Kintner, Hirsch and Deutch."

DOE was officially formed in October 1977. Many of the programs that had reported to Hirsch, including fusion, were now in an Office of Energy Research, headed by John Deutch, who had been recruited by Secretary of Energy James Schlesinger. Deutch was a professor of chemistry at MIT at the time. Deutch, as Schlesinger before him, would later serve as head of the CIA. When President Carter's first budget (FY 1979) was submitted to Congress in early 1978, it included a proposed $60 million cut for the US fusion program.

Deutch set up a panel with the aim of getting justification for cutting the fusion program budget. For panel chairman, he chose John Foster, a vice president at TRW and a former director of the Livermore laboratory. Rather than endorsing fusion budget cuts, however, the Foster panel validated the accomplishments and promise of the fusion program but argued that too much emphasis was being placed on the tokamak. The current emphasis on tokamaks was the result of a "bandwagon effect" they said. Though recognizing the success of the tokamak, they said, "While tokamaks are currently the most advanced scientifically, they seem on the surface to be the most complex of the possible alternative approaches to fusion from the standpoint of engineering into an energy producer." Rather than a demonstration fusion power plant around the year 2000, as had been put forth in the 1976 fusion plan, the Foster panel suggested 2015 for a demo and "full commercialization" by 2050. Most importantly, the Foster committee recommended that the fusion budget not be cut [4]. Although the Foster committee report is sometimes cited [4] as deviating from the 1976 plan by objecting to the emphasis on tokamaks, in fact it differed in only one respect: the timescale for the demo. The 1976 plan acknowledged the tokamak to be in the lead, but it was not to be "the only horse in the race." Indeed, the 1976 plan provided for a number of competing concepts, including mirrors and theta pinches.

Based on 2XIIB, having achieved its promised goals in 1975, I recommended in April 1976 and got approval for Livermore to construct a much larger superconducting version of 2XIIB [28]. We first named it MX. It aimed for operation in 1981. Although it would not burn tritium, it would, in many respects, put the mirror program on a track to have some hope of competing with the tokamak.

A mirror fusion reactor based on scale-up of 2XIIB geometry, even assuming success in reducing end losses, still did not solve the issue that it likely could barely produce breakeven. A commercial power plant needed to do better than that, though other applications like materials testing or hybrid fusion–fission would still be possible. Attention now began to turn to how one might proceed beyond 2XIIB (or MX) to a more reactor-relevant mirror configuration. There were frequent discussions between my staff (Arthur Sleeper, Bill Ellis, Milt Johnson) and Livermore and other scientists about this issue, which went by names like "end plugging" and "Q enhancement" ($Q=1$ is the definition of breakeven). Livermore got the message: 2XIIB success would not be enough to keep mirrors in the race with tokamaks. In the panel I had chaired in early 1976, several end-plugging possibilities had been identified [28]. The eventual solution of choice, however, was not on that early list.

The "solution" put forth by Ken Fowler, David Baldwin, and, independently, Grant Logan [29] came in the form of the "tandem mirror." My vision for this was a central mirror cell with a 2XIIB mirror plasma at either end serving as "end plugs." Their vision of the end plugs was considerably more complex. To test the general tandem idea, we authorized Livermore to construct a "small" Tandem Mirror Experiment (TMX) to replace the existing Baseball facility. The TMX (Fig. 4.6) was authorized in January 1977 and began operation in October 1978. It had a neutral beam-sustained plasma on either end. TMX quickly validated the basic physics of the tandem mirror concept; consequently, we authorized upgrading the planned MX from a single cell to a tandem configuration, renaming it Mirror Fusion Test Facility (MFTF). MFTF, of course, would be much more ambitious and expensive (over $200 million) than MX and take longer to build (new completion date 1985). Its schedule, however, would be consistent with the longer overall fusion development schedule recommended by the Foster panel.

Though the TMX/MFTF style of end plugs might successfully demonstrate a significant degree of Q enhancement, even more was desired for a power plant. In 1979, as I was preparing to leave government to form Fusion Power Associates, the Livermore scientists (primarily Fowler and Logan) designed more complicated end cells (called thermal barriers). To test this idea, they would have to rebuild the TMX end cells. Kintner approved the new device, called TMX-Upgrade, or TMX-U, but it would not operate until 1983. This was "too long to wait" in Fowler's mind. He wanted to modify the original MFTF proposal to have end cells based on the new (but untested) thermal barrier tandem idea. This was a high-risk, high-payoff approach. Livermore laid the new proposed device (renamed MFTF-B) on Ed Kintner's desk in late 1979 and it was subsequently approved.

Power Plant Designs

As our more goal-oriented fusion program took shape in the 1973–1975 period, Bussard and Jim Williams (who succeeded Bussard upon his departure in 1974) initiated a serious effort aimed at conceptual designs of power-producing fusion devices: engineering test facilities, experimental power reactors, and demonstration power plants. A group formed by Harold Forsen at the University of Wisconsin that included Bob Conn and Jerry Kulcinski turned out a series of tokamak power plant designs called UWMAK-I, UWMAK-II, and UWMAK-III in 1973, 1975, and 1976, respectively. Today's tokamak power plant designs do not look a lot different, at first glance, than those depicted in those studies. Groups led by Bill Stacey (Argonne National Laboratory), Charlie Baker (General Atomics), and Mike Roberts and Don Steiner (Oak Ridge) undertook designs of experimental power reactors and engineering test reactors [30].

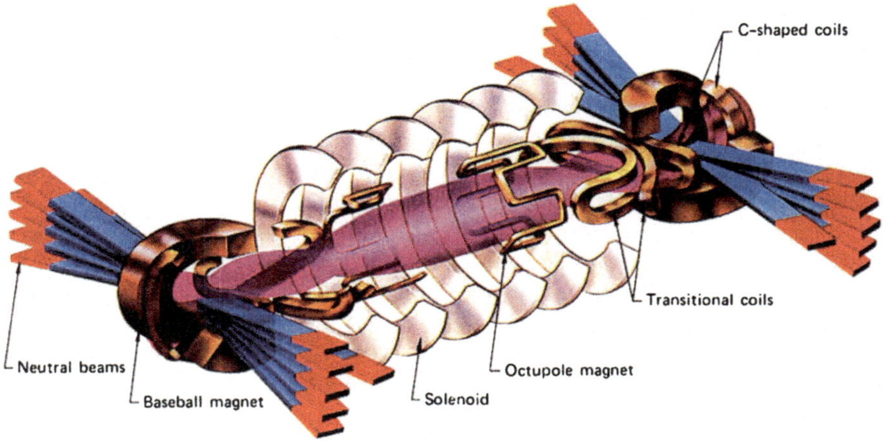

Fig. 4.6 Schematic of Tandem Mirror Experiment TMX (Lawrence Livermore National Laboratory)

In 1978, the International Atomic Energy Agency (IAEA) of the United Nations began a coordinated multinational design effort of a tokamak to be built in the 1980s to produce substantial net fusion power. It was given the acronym INTOR (for International Tokamak Reactor). Early members of the US INTOR team included team leader Bill Stacey, (now at Georgia Tech), Jerry Kulcinski, Paul Rutherford (Princeton), and John Gilleland (General Atomics). The initial INTOR steering committee consisted of Stacey, Guenter Greiger (European Commission), Boris Kadomtsev (USSR), and Segeru Mori (Japan). A large team was assembled among the various parties. By October 1979, the group had completed a 650-page design report for the INTOR and estimated its cost at between \$1.5 and \$2.3 billion. They were ready to proceed into engineering design and construction. The understanding during the INTOR study was that the design was available for any country to construct or that partnerships could be formed. Within the USA, Kintner wanted to proceed with a US project [30]. Thus, the schedule for building a US (or international) engineering test reactor was still on schedule for a decision in accordance with the plans we had prepared in 1972 and 1976.

Surpassing the Lawson Ideal Ignition Temperature in a Tokamak

The PLT began plasma operations in 1974, as scheduled; the beams were all delivered from Oak Ridge a few years later, and four powerful beams were in full operation on PLT in 1978. In the summer of 1978, temperatures in excess of Lawson "ideal ignition" value of 5 keV were achieved. We all considered this to be the most significant achievement in the nearly 30 year history of fusion research and to be the achievement of milestone 1 of the 3-milestone development strategy laid out in

1973. A press release was hastily prepared at Princeton and approved (but not released) at DOE.

Energy Secretary James Schlesinger had a passion for being in control of the news (a common trait among high-level government officials). He had his own priorities for what should be making news, and fusion was not on his list. He was pushing conservation and coal projects at the time. His office sat on the fusion news release for weeks. Eventually, as often happens, the news media got wind of the Princeton results and, on Saturday August 12, a story was running on the news wires that something important had happened at Princeton. I was at home when my phone rang. It was the DOE Public Affairs officer on duty. She said that they were getting calls from news organizations for comment and that they could not reach the Secretary of Energy's press officer, Jim Bishop, or other high-level officials. Could they refer the calls to me, they asked. As I began taking calls, it was clear that much of the story was already out. I determined that the best course of action was just to tell the truth. My view was that "stonewalling" the press would only complicate matters. Mel Gottlieb from Princeton called me shortly after I began taking calls to say that he understood that the press release was still embargoed at DOE and he asked me what he should do. I told him the "cat was already out of the bag" and that he should just "come clean." Still, I did not expect that it would be a really big story in the grand scheme of national news, even though it was a "big deal" for fusion. So, when I opened my front door on Sunday morning to retrieve my Washington Post paper [31], I was shocked to see a front-page banner headline "US Makes Major Advance in Nuclear Fusion." Further down the front page, below the fold, was an article titled "Pope is Buried."

The fusion article, by Post staff writer Bill Peterson, noted that "Princeton University scientists have made a major advance in the race to tame nuclear fusion" by producing "sun-like temperatures of more than 60 million degrees." It quotes me as saying, "It is the first time we've produced the actual conditions of a fusion reactor in a scale-model device" and saying, "This is the biggest thing that has happened in fusion research." It also quotes me as saying "The question of whether fusion is feasible from a scientific point of view has now been answered. The practical questions of what price it will cost and when it will be duplicated commercially are now engineering and economic questions." The article noted that one glass of water contains fusion fuel equal to the energy of a ton of gasoline and that one pound of heavy hydrogen produces the same energy as 5,000 tons of coal. It quotes Princeton's Harold Furth as saying "All I can say is we've developed something really important. Nothing has happened of this significance in fusion research in the last 10 years."

What happened next was a classic example of Washington political bureaucracy in action. Late Sunday night I got a call saying that Kintner and I were to report to Schlesinger's office early the next morning. When we got there, we were told to wait outside while Schlesinger's assistant secretary John Deutch and others were consulting on the matter of how to do "damage control." We later learned that Schlesinger had told Deutch "fire Kintner, fire Dean." He believed we had deliberately leaked the embargoed data to the press in order to buttress our case for bigger budgets. We also learned that Schlesinger's press officer, Jim Bishop, had called Gottlieb at Princeton and told him to put out a statement saying the results were not that significant. Gottlieb refused. He advised William Bowen, the president of Princeton University of the situation. Bowen then called Schlesinger to say that Princeton would not back down from the story and that if DOE put out a statement that the results were not significant, then Princeton would put out a story contradicting the DOE. Schlesinger backed down. In the end, Deutch held a press conference that same Monday afternoon, with Gottlieb present, saying what an important event this was scientifically but that fusion was not "just around the corner." (The Post article had ended with "It will probably be 20 to 30 years before the commercial applicability of fusion reactors is demonstrated, Dean said.") Kintner and I were not fired. Kintner, in fact, had had no part in the whole episode, since he was on a plane returning from the West Coast on Saturday when the story broke. Deutch's deputy, Eric Willis, told me that he "hoped I had learned my lesson of not talking to the press." But the lesson I had learned (once more) was that when the technical people stand firm, the bureaucrats back down. Author Tom Heppenheimer provides a much lengthier and colorful description of these events [32]. He says, for example, that as the DOE press conference was about to begin, Deutch told Kintner, "Go sit in the back of the room and shut up."

During the 1970s, the general public often got their first exposure to fusion at US airports, where a cadre of enthusiastic followers of perennial presidential candidate Lyndon LaRouche proselytized and solicited passengers and passersby concerning ideas for the future of society. Distribution of a magazine called "Fusion" was part of their pitch. A LaRouche-affiliated group called the Fusion Energy Foundation published the magazine. Large-scale deployment of fusion power plants was integral to LaRouche's vision for society. The articles in Fusion magazine were well written and largely accurate. Key contributors included writers Chuck Stevens and Marsha Freeman. They frequently attended fusion conferences. Fusion researchers were generally reluctant to associate too closely with them because of their affiliation with LaRouche. Nevertheless, they did have some political impact, though the degree is hard to measure. They are often credited, for example, for having "leaked" the PLT story to a Miami newspaper that initiated the events described above. Fusion magazine has been succeeded by a new magazine called 21st Century Science and Technology, edited by Marjory Mazel Hecht. Marsha Freeman still writes for the magazine and in 2009 published an excellent article on the Korean fusion program [33].

As the decade of the 1970s was coming to a close, everything seemed to be in place and on schedule to implement the 1976 plan. The Confinement Systems Division for which I had been responsible had accomplished its goals for the decade: in the tokamak program, PLT had surpassed the Lawson ideal ignition temperature, the TFTR was under construction, and several "medium-size" tokamaks were in operation, including Alcator (MIT), PDX (Princeton), and Doublet III (General Atomics). In the magnetic mirror area, the MFTF had been approved; TMX was in operation and providing validation of the tandem concept. In the high-density systems area, the Scyllac experiment had been terminated; the Z-pinch program had been transferred to the Division of Applied Plasma Physics; a new bumpy torus "Proof of Principle" device, EBT-P, had been proposed by Oak Ridge; and the lab had agreed to put the project out for bid to industry.

Fusion Power Associates

I decided in early 1979 to leave government. I felt it was time to "mobilize the private sector" to do the engineering development leading to a commercial fusion power plant. After several discussions with industry executives, I held a "pre-incorporation meeting" on April 2, 1979, at the Nassau Inn in Princeton, NJ, attended by 23 people from 20 organizations. We agreed to form an association called Fusion Power Associates (FPA) and agreed on the bylaws and membership fees. Alvin Trivelpiece and Nicholas Krall had since left the University of Maryland to become vice presidents of Science Applications, Inc. (SAI), in La Jolla, CA. I was hired by SAI and the company provided legal and financial assistance for start-up. FPA was incorporated as a nonprofit research and educational foundation in the State of California on August 16, 1979. Trivelpiece, Krall, and myself signed the articles of incorporation. One year later, we were granted 501(c)(3) nonprofit, tax-exempt status by the Internal Revue Service. The first board of directors meeting was held on September 24, 1979, in La Jolla to approve the bylaws, etc. The second board meeting was held on October 14, 1979, in Albuquerque, NM, to approve the applications of ten companies (later extended to thirteen) to become "charter members" of FPA. They are listed in Table 4.2. The FPA bylaws provided that not all members of the board would be from the member institutions. The members of the first board and their affiliations are shown in Table 4.3.

A press release announcing the formation of Fusion Power Associates was issued on November 9, 1979. It stated the FPA goals as:

1. To help bridge the gap between scientific research on fusion energy and its engineering development
2. To assist in the development of increased public awareness and understanding of the potential of fusion energy
3. To foster cooperation in fusion research and development among all public and private organizations, including government, universities, laboratories, and industry

In a letter to Fusion Power Associates dated December 3, 1979, from Vice President Walter Mondale's Office, his Special Assistant for Domestic Policy, Eric Vaughn, wrote, "Magnetic fusion programs are moving along very well, in our opinion. So well, in fact, that the Department of Energy and the Office of Management and Budget think it is time to move magnetic fusion out of the pure research stage." He further stated, "Inertial confinement fusion is also moving ahead at a pace considered by many to be slow by comparison to the magnetic fusion pace, but encouraging nonetheless."

During 1979, Congressman Mike McCormack (D-WA), with the help of his energetic staff member, Al Mense, was considering how to move the fusion program forward expeditiously toward fusion power. He formed an advisory committee

Table 4.2 Charter members of fusion power associates

BDM Corporation, McLean, VA	Mathematical Sciences Northwest, Bellevue, WA
Burns and Roe, Inc.	McDonnell–Douglas Corporation, St. Louis, MO
Ebasco Services, New York, NY	Thermo-Electron Corporation
General Atomic Corporation, San Diego, CA	Quadrex Corporation, San Jose, CA
ILC Technology, Sunnyvale, CA	Science Applications, Inc., La Jolla, CA
JAYCOR, Del Mar, CA	Westinghouse Corporation
KMS Fusion, Inc., Ann Arbor, MI	

Table 4.3 First board of directors of fusion power associates

Stephen O. Dean, Chairman Henry J. Gomberg, Vice Chairman (KMS Fusion, Inc.) Donald L. Kummer, Secretary (McDonnell–Douglas Astronautics Co.) Bernard J. Eastlund, Treasurer (BDM Corp.) Ronald C. Davidson (MIT) Nicholas A. Krall (JAYCOR) Sherman Naymark (Quadrex Corp.) Tihiro Ohkawa (General Atomic Co.)	Paul Reardon (Princeton University) Leonard Reichle (Ebasco Services) Peter H. Rose (Mathematical Sciences Northwest) Glen Sorenson (ILC Technology, Inc.) Alvin W. Trivelpiece (Science Applications Inc.) James M. Williams (Los Alamos Laboratory) Gerold Yonas (Sandia Laboratories) Donald P. Zeifang (National Assoc. of Broadcasters)

Table 4.4 Members of Rep. Mike McCormack's 1980 Fusion Advisory Committee

Robert L. Hirsch (Exxon), Chairman	Joseph Gavin (Grumman)
Richard Balzhiser (EPRI)	John Landis (Stone and Webster)
Robert Conn (U. Wisconsin)	Tihiro Ohkawa (General Atomics)
Ersal Evans (Westinghouse)	Robert Smith (Public Service Electric and Gas of NJ)
T. K. Fowler (Lawrence Livermore Lab)	Alvin Trivelpiece (Science Applications, Inc.)
Harold Furth (Princeton Plasma Physics Lab)	

chaired by Bob Hirsch (then at Exxon) which began meeting that summer. The members of the committee are shown in Table 4.4.

On the afternoon of December 10, 1979, the Hirsch committee met in a Hearing Room of the House of Representatives Science and Technology Committee, of which McCormack was chair. FPA held a reception in the Hearing Room after their meeting. McCormack's Congressional Committee held a hearing on the fusion program the next day.

The Carter Plan vs. The Reagan Agenda: 1980–1985

5

I believe it is obvious that once fusion is achieved, the energy shortage will be past and we will be independent of foreign fuels.

Ansel Adams, Photographer

The Magnetic Fusion Energy Engineering Act of 1980

On January 21, 1980, Congressman Mike McCormack sent a formal letter to President Jimmy Carter requesting he establish "as a national goal" the operation of a fusion electric demonstration plant before the end of the century. At the Department of Energy, Ed Frieman, previously the deputy director of the fusion program at Princeton University Plasma Physics Laboratory, was confirmed as Director of Energy Research. Frieman set up a new fusion policy review panel chaired by Solomon J. Buschbaum (Bell Laboratories) to update the 1978 DOE review, chaired by John Foster, for John Deutch. Members of the Buchsbaum committee are shown in Table 5.1.

Congressman McCormack, with 135 cosponsors, introduced a Bill (H.R. 6308) declaring (if passed) it "to be the policy of the United States" to operate a fusion demonstration plant "before the end of the twentieth century." At a hearing on this Bill on March 6, I testified, "In view of the impressive progress of the 1970s and the encouraging outlook for the successful development of practical fusion energy systems," I recommend that the U.S. government adopt the following policies:

- Establish, as a national goal, the operation of a practical fusion energy demonstration plant by the year 2000.
- Accelerate the present national effort over the next few years and maintain that effort at the level required to meet the national goal.
- Commit immediately to a program leading to construction and operation of a fusion engineering test facility based upon the tokamak concept.
- Continue the commitment to search for concepts which show promise for practical commercial fusion systems and to rapid development of the most promising.

S.O. Dean, *Search for the Ultimate Energy Source: A History of the U.S. Fusion Energy Program*, Green Energy and Technology, DOI 10.1007/978-1-4614-6037-4_5,

Table 5.1 Members of 1980 Buchsbaum Fusion Review Committee

John Foster (TRW)	James Fletcher (University of Pittsburgh)
Eugene Fubini (Fubini Consultants, Inc.)	Wolfgang Panofsky (Stanford Accelerator Center)
Marshall Rosenbluth (Inst. for Adv.. Study)	Robert Conn (UCLA)
Marvin Goldberger (Calif. Inst. of Technology)	Roy Gould (California Institute of Technology)

- Strengthen those aspects of inertial confinement fusion dealing with civilian applications.
- Commit to the engineering development of the technology required to meet facility operation schedules.
- Establish policies which encourage the full participation of industry in all phases of fusion development.
- Continue to foster cooperation on fusion research among all nations for the ultimate benefit of mankind.

I stated, "The U.S. government should be commended for making a sustained investment in fusion research for the past 28 years. The time has come to begin to capitalize on this investment by placing increased emphasis during the 1980s on the engineering and systems design aspects of practical fusion energy systems."

On April 22, President Carter responded to Congressman McCormack's January 21 letter saying, "I am aware of the promise that fusion energy holds for long range, relatively clean, and inexhaustible energy. I strongly support the development of a technology that offers such hope for meeting future energy needs." He added, "The Administration is committed to the fusion option."

In June, the Hirsch committee and the Buchsbaum committee issued their reports. Both groups focused their reports on the readiness of the fusion program to proceed with fusion engineering development, including the construction of a fusion engineering test facility (ETF) during the 1980s. The Hirsch committee told Congressman McCormack that they reaffirmed their previous endorsements of the readiness of the program for engineering development. The Buchsbaum committee provided its report to the DOE's Energy Research Advisory Board (ERAB). They stated, "The magnetic fusion program can, and should, embark on the next logical phase towards its goal of achieving economic feasibility of magnetic fusion. To this end, a broad program of engineering experimentation and analysis should be undertaken under the aegis of a Center for Fusion Engineering (CFE)."

In July, Senator Paul Tsongas (D-MA), with six other cosponsors, introduced legislation (S-2926) in the Senate similar to that of Rep. McCormack's in the House. In August, the US House of Representatives passed H.R. 6308 by a vote of 365–7. In August also, the DOE's ERAB endorsed the Buchsbaum committee report, sending it on to Secretary of Energy Charles Duncan, urging him to institute "a broad program of (fusion) engineering experimentation and analysis . . . under the aegis of a Center for Fusion Engineering." The ERAB told Duncan the construction of a fusion engineering device would require a doubling of the fusion program budget in 5–7 years.

The House and Senate versions of the fusion Bills were quickly merged, approved by voice votes with strong bipartisan support in the two chambers, and sent to the President. President Carter signed into law the Magnetic Fusion Energy Engineering Act of 1980 [34] on October 7, 1980. The law states, "The Secretary (of Energy) shall initiate design activities on a fusion engineering device using the best available confinement concept to ensure the operation of such a device at the earliest practicable time, but not later than the year 1990." The law further states, "The Secretary shall initiate at the earliest practical time each activity which he deems necessary to achieve the national goal for operation of a commercial demonstration plant at the turn of the twenty-first century."

A US presidential election was under way at the time. One month after he signed the Fusion Engineering Act, President Carter lost his reelection bid to Ronald Reagan.

Fusion Power Associates held its first annual meeting and symposium, *The Status of Fusion R&D and The Role of Industry in Fusion Development* in November 1980 in Washington, DC. At the meeting, FPA presented its first Leadership Awards to Rep. Mike McCormack, Sen. Paul Tsongas, Robert L. Hirsch, and Solomon J. Buchsbaum. Speaking at the symposium, Rep. McCormack said, "the formation of Fusion Power Associates and the involvement of industry now are essential elements in the coming engineering development phase which will lead to fusion power." McCormack, unfortunately, had not been successful in his own reelection to Congress for the next term. The proceedings of the symposia were published in book form [35].

The Pewitt Problem

During the transition from President Carter to President Reagan, the DOE Director of Energy Research (a presidential appointment position) Ed Frieman learned that he would not be retained. His deputy, Douglas Pewitt, became acting director until a new presidential appointee would be named and confirmed. Before joining DOE, Pewitt had been a budget examiner at the Office of Management and Budget (OMB). Pewitt and Kintner would soon clash. At a February 25, 1981, House Science and Technology Committee hearing on fusion, Pewitt stated, "The Department cannot embark upon a program that is predicted on an unbroken series of technical successes; we must plan prudently and incorporate into our planning the possibility for scientific disappointments or setbacks." He said that several ongoing projects (unspecified) would likely be cancelled, saying, "none of these projects is absolutely essential." After hearing about an unflattering comment on his testimony attributed to Kintner's deputy, John Clarke, Pewitt reassigned Clarke to do a 6-month study, beginning in March 1981, that had nothing to do with fusion. He did this without consulting Kintner.

One of the mandates of the Magnetic Fusion Energy Engineering Act of 1980 was "The Secretary of Energy shall develop a plan for the creation of a national magnetic fusion engineering center for the purpose of accelerating fusion technology development via the concentration and coordination of major magnetic fusion engineering devices and associated activities at such a center." It also stated that it was the intent of the Act "to promote the broad participation of domestic industry in the national magnetic fusion program." The Act required the Secretary of Energy to submit a report not later than July 1, 1981, "setting forth the steps necessary for implementation" of the establishment of the engineering center (dubbed the Center for Fusion Engineering or CFE).

During the first half of 1981, Pewitt prepared the CFE report for Secretary James Edwards, who transmitted it to Congress on July 7. It stated, "We have determined that it is premature to establish fully the national magnetic fusion engineering center at this time." Instead, DOE proposed to set up an Engineering Feasibility Preparations Project (EFPP) at an existing site. This report did not sit well with the chair of the House Science and Technology Committee, Rep. Don Fuqua, or with Rep. Marilyn Bouquard who had replaced Mike McCormack as chair of its Subcommittee. Rep. Fuqua, in response to the Secretary, said, "I would ask that you reconsider your present intention to establish an EFPP and instead move forward, within your 1982 budget, with formally soliciting proposals from all interested parties on establishing a CFE management scheme and operating team." Rep. Bouquard wrote the Secretary, "The proposal (to establish an EFPP) smacks of trying to undermine the Congressional idea of CFE while adding yet another laboratory as a player in a complex institutional program. Congress has clearly told you they want appropriate responsibility shifted to industry."

President Reagan had campaigned on a Republican Party platform that called for the abolishment of the DOE. He vigorously wanted to eliminate much of the energy industry regulations and energy development programs set up in the Carter Administration. All federal budgets were under scrutiny for possible reductions, especially the DOE. In the ensuing OMB-led "scrubbing" of the Carter budgets, those programs closer to commercialization (like fossil and solar energy) took the biggest hits. Those farther from commercialization, like fusion, fared better. The fusion budget, in fact, would grow in FY 1982 (the first Reagan budget) compared to the FY 1981 (last Carter) budget, but not to the levels required to implement the Fusion Engineering Act.

Kintner Resigns

Ed Kintner did not like or trust his new temporary boss, Douglas Pewitt. The saga of their interactions and the aftermath is treated in some detail by Heppenheimer [36]. Though Pewitt only stayed at DOE during the first 6 months of 1981, he continued to plague Kintner psychologically. In the summer of 1981, Al Trivelpiece (cofounder of Fusion Power Associates) was named DOE Director of Energy Research, and Pewitt left DOE to work for the President's Science Advisor, George Keyworth at OSTP. OSTP and OMB usually worked hand-in-glove on science and technology budgets. In September, the fusion budget examiner, Dom

Repici, left OMB. He had been a staunch supporter of an aggressive fusion development strategy. His boss, Tom Palmieri, was also supportive but perhaps less versed in the details.

Trivelpiece quickly arranged with the new Energy Secretary, James Edwards, to move ahead with a competitive procurement to industry for the fusion engineering device design and the Center for Fusion Engineering. But then President Reagan ordered a 12% reduction in planned FY 1982 federal spending. During the summer of 1981, The Office of Management and Budget (OMB) was carrying out these instructions from President Reagan to cut FY 1982 planned spending and was also simultaneously reviewing DOE's FY 1983 budget requests. OMB did not want to cut the FY 1982 fusion allocation by 12%; they preferred to apply the cut to DOE as a whole and distribute the cuts unequally among the DOE programs. They did decide, however, to propose that the fusion budget be spent differently than DOE (i.e., Kintner) had planned. They did this on the basis of a consultant's recommendation that funding for the construction of the magnetic mirror MFTF-B be reduced and the money spent on other fusion research programs. Kintner did not like either the consultant or his recommendations. He especially did not like the idea that OMB could tell him how to spend his allocation. Ed Kintner believed strongly in keeping up a competition between the Mirror and Tokamak approaches. He was committed to keeping the MFTF-B construction project on schedule.

Kintner believed that Pewitt was at least partly responsible for the OMB "meddling" in his prerogatives as head of the fusion program (this may or may not have been the case). In a November 9, 1981, meeting among Kintner, Trivelpiece, OMB budget examiner Tom Palmieri, and, his consultant, Steve Bodner, they tried to resolve the FY 1982 fusion allocation issue. Over the next week, Trivelpiece tried to come up with a compromise that would satisfy all the parties. In the end (while Kintner was overseas at a fusion meeting), Trivelpiece and OMB agreed to cut the MFTF-B construction budget by $25 million (originally OMB wanted to cut it $42 million) and to reallocate that money to other fusion research programs. It had the effect that fusion was exempted from the FY 1982 Reagan-ordered reductions. This satisfied everyone but Kintner. It was a matter of principle to him that no one, especially no one at OMB, was going to tell him how to allocate the fusion monies among its subprograms. He had worked for Admiral Rickover, who never allowed anyone else in the Executive Branch or Congress to tell him how to run his program. Kintner felt likewise. In his mind, if he gave in on this point, he had lost control of his program. He also had come to believe that the new Administration had no intention of constructing a fusion engineering test reactor in the near future or to implement the Magnetic Fusion Energy Engineering Act of 1980, at least not on the schedule, or with the funding, the Act demanded. Indeed, the President's Science Advisor, George Keyworth, had stated that the government would build no more "flagship" energy demonstration facilities. The development of new energy sources would henceforth be the responsibility of the private sector, he said.

Kintner resigned effective January 1, 1982. In his letter of resignation, an edited copy of which is reprinted in Heppenheimer's book [36], Kintner said in part, "the ability of Program Directors like myself to maintain program direction and cohesion has been badly eroded." He said, "The ultimate example is the recent OMB-OSTP action on the FY 1982 budget. At worst it is malicious meddling; at best it is ignorance of the broader implications of the action." Kintner also said, "it now appears that the goal-oriented magnetic fusion program which has been put together over the past 5 years will not be carried out."

The end result of this first year (1981) of the Reagan presidency was that the budgets and schedule called for in the Fusion Engineering Act had been delayed but not necessarily abandoned. The Administration had indeed put the fusion engineering test facility on a delayed schedule but had not abandoned it altogether as a potentially justifiable government-funded project at some point. The FY 1983 fusion budget proposed by President Reagan in January 1982 was $444 million (compared to $394 million in the FY 1981 Carter budget). In the end, Congress would appropriate $461 million for magnetic fusion for FY 1983. While a victory of sorts, the FY 1983 proposed budget was well short of the $596 million envisaged in the Act.

When Kintner left DOE, I offered, and he accepted, office space at Fusion Power Associates. He had many contacts in the nuclear industry, and it was my hope that he would help me to interest industry in joining FPA and urging implementation of the McCormack/Tsongas Fusion Engineering Act. In that, I was to be disappointed. When Ed resigned from his post as head of the DOE fusion office, he determined for the most part to walk away from fusion. He did participate in an FPA fusion symposium later that year, and he published an article in the MIT Technology Review entitled "Casting Fusion Adrift" [37]. In that article, he said, "The nation's pace-setting program for exploiting the ultimate energy source is in danger of going under." He stated, "Recent implicit and explicit decisions have dimmed the prospects that the strategy and timetable mandated by the (Magnetic Fusion) Energy Engineering Act can be carried out." He noted that the Act (and he) had envisaged a FY 1982 budget of $525 million and a FY 1983 budget of $596 million. He stated that he had requested $557 million for FY 1983 as an amount that could "balance the need to reduce expenditures with the strong program recommendations made by the review panels and enacted by Congress." He said that the $444 million requested by the President for FY 1983, at the recommendation of the OMB, "does not provide for the broadening of activities recommended by the Foster committee nor the engineering initiatives recommended by the Buchsbaum panel and authorized by Congress." He said the result is to "leave the fusion program without a strategic backbone—it is a collection of individual projects and activities without a defined mission or timetable." He said, "The plan to increase industry involvement in fusion development is postponed indefinitely, and the industrial and economic benefits of high-technology spinoffs, surely an increasingly important by-product of an accelerated fusion technology program, will be lost."

Fusion Power Associates Board of Directors presented its 1981 Leadership Award to Edwin E. Kintner.

In March 1979, the Three Mile Island nuclear power plant in Harrisburg, PA, had suffered a meltdown. The plant was owned by GPU Nuclear with headquarters in New Jersey. The head of GPU Nuclear in 1982 was an old Rickover colleague of Ed Kintner and convinced Ed in 1983 to take a job in New Jersey as an executive Vice President of GPU Nuclear to oversee the Three Mile Island cleanup. Kintner would soon be dissatisfied with the cleanup progress, which was being done by a team led by Bechtel Corporation. Kintner wanted his own man on-site to ride herd on the Bechtel project manager. He offered me the job. I went to Harrisburg with Ed for a management meeting there. I considered the job but, although getting Three Mile Island cleaned up was clearly important, overseeing the cleanup seemed to me a thankless task and one for which I had no particular aptitude or interest compared to fusion. So I turned it down.

After Kintner left DOE in January 1982, DOE Director of Energy Research Al Trivelpiece put his own deputy, Jim Kane, in charge of the fusion program in an acting capacity while he considered Kintner's replacement. I was sounded out as to whether or not I wanted to be considered. If it had appeared that the Administration was seriously considering implementing a goal-oriented fusion engineering effort, I would have enthusiastically put my hat in the ring. I had spent 17 years in government service and was at the top of the government salary scale when I left. The first of my three children had just entered college. I was finally making enough money that I could see my way to being able to pay for their college education. I was only 2 years into building up Fusion Power Associates, and the thought of going back to DOE and fighting a budget-stingy bureaucracy did not excite me. Although I was not offered the job nor even pressed hard to consider it, I asked not to be considered.

In March 1982, Trivelpiece announced his decision to name John Clarke to head the fusion office. He was eminently qualified, having been Kintner's deputy since 1978, a former head of the fusion program at the Oak Ridge lab, and a cochair of my 1973 tokamak assessment report [26]. Since 1980, Clarke had been the driving force for organizing the community on the design and planning for construction of a tokamak fusion engineering device (FED). Trivelpiece also moved Jim Decker from his position as Director of the Applied Plasma Physics Division of the Office of Fusion Energy to be his special assistant. Jim would later become Director of the Scientific Computing Program and Principal Deputy Director of the Office of Energy Research under Trivelpiece. Over the next two-plus decades, Jim Decker would often find himself in the position of being "acting director" of the Office of Energy Research (later renamed Office of Science) at DOE, when current directors left or were being replaced in new administrations. He was a godsend of stability on many such occasions.

In February 1982, the Electric Power Research Institute (EPRI) published the results of a study they had sponsored, "Utility Requirements for Fusion" that had been prepared by Burns and Roe, Inc. Noel Amherd, from the EPRI Advanced Power Systems Division, was project manager for the study. The study resulted from a survey of 43 utilities. The report [38] states, "in addition to considering the

Table 5.2 Utility requirements for fusion (from [38])

A. Utility Planning and Finance	C. Utility Operations
1. Plant Capital Cost	1. Plant Operating Requirements
2. Plant Operating and Maint. and Fuel Costs	2. Plant Maintenance Requirements
3. Forced Outage Rate	3. Electrical Performance
4. Planned Outage Rate	4. Capability for Load Change
5. Plant Life	5. Part Load Efficiency
6. Plant Construction Time	6. Minimum Load
7. Financial Liability	7. Startup Power Requirements
8. Unit Rating	
B. Safety, Siting and Licensing	D. Manufacturing and Resources
1. Plant Efficiency	1. Hardware Material Availability
2. Plant Safety	2. Industrial Base
3. Dependence on Other Systems	3. Natural Resource Requirements
4. Flexibility of Siting	4. Fuel Availability
5. Waste Handling and Disposal	
6. Decommissioning	
7. Licensability	
8. Weapons Proliferation	

cost of energy for a power plant, utilities consider 23 other requirements," of which four were considered "vital to plant acceptability: Plant Capital Cost, Financial Liability, Plant Safety, and Liability." The full list of requirements is shown in Table 5.2.

Mirrors and Bumpy Torus

During 1979, Livermore scientists Ken Fowler, David Baldwin, and Grant Logan were developing a theoretical model for a better tandem mirror design, using end cells called "thermal barriers." To test the ideas, it was proposed and approved to modify the TMX experiment and rename it TMX-U. Livermore also proposed to modify the design of the tandem version of MFTF to include thermal barrier end cells and to rename it MFTF-B.

Magnetic Mirror Systems Branch Chief Bill Ellis convened a Mirror Senior Review Panel to provide "An Assessment of the U.S. Mirror Fusion Program" [39]. I served on that panel. A separate review had been conducted for the Bumpy Torus (EBT-P) project. On February 13, 1980, the Fusion Power Coordinating Committee (FPCC), which Kintner chaired, provided a "position statement" on the MFTF-B and EBT-P projects. They noted that both the EBT-P and the original MFTF projects had previously been approved and that funding for initiating them had been provided in the FY 1980 and FY 1981 budgets. They stated, "The FPCC endorses both projects and recommends that the required additional funding be sought to keep these projects as near to the proposed optimum schedules as possible." Kintner approved the newly defined MFTF-B thermal barrier version, and its construction proceeded into the early 1980s. Kintner subsequently split my Confinement Systems Division into two divisions: a Tokamak Systems Division (headed by N. Anne Davies) and a Mirror Systems Division (headed by Bill Ellis). The latter assumed responsibility for the Bumpy Torus. The High Density Systems Branch was eliminated in the process.

Inertial Confinement

Inertial confinement fusion (ICF) made impressive progress during the 1970s. New, larger lasers had been built and operated at the Livermore lab, pulsed power facilities for beam-driven ICF were in operation at the Sandia lab in Albuquerque, and heavy ion-driven inertial fusion was being pursued at the Lawrence Berkeley lab. The dominant funding source for these activities was the DOE Defense Program. The energy application of ICF was motivating the researchers but not the funding managers.

Fusion Power Associates held its second annual meeting and symposium on December 10–11, 1981, in San Francisco on the theme "Status of the World Effort on Inertial Confinement Fusion." Among the attendees were Geoffrey Manning (Director of the UK's Rutherford Laboratory), N. G. Basov (Director of the USSR's Lebedev Institute and coinventor of the laser), and Chiyoe Yamanaka (Director of Japan's Institute of Laser Engineering at Osaka University). Following the meeting, the Fusion Power Associates Board of Directors issued the following statement titled "Inertial Confinement Fusion—Energy Applications":

> "Congressional and DOE attitudes have resulted in considerable pressure on ICF program managers to eliminate studies aimed at energy applications of inertial fusion. Fusion Power Associates believes that to respond to these pressures by terminating the minimal ongoing studies of potential energy applications of ICF is not in the national interest. Military research and development have historically led to many important contributions to the civilian sector and to the economy. Government sponsors should be proud of, and encourage, these connections. Permitting a reasonable percentage of the ICF funding to continue to be spent on energy applications is the best way to ensure that the nation ultimately derives full benefit from the ICF program."

Management

The management structure established by Bob Hirsch that empowered the Washington fusion office to direct the US fusion program had at its heart the Fusion Power Coordinating Committee (FPCC). With the DOE fusion director as its chair, his division directors, and the main laboratory (and General Atomics) project directors as members, agreement with a proposed course of action by the FPCC constituted an endorsement that carried substantial weight with "higher level" managers at DOE and with OMB and Congress. This structure was abolished by Clarke and Trivelpiece as soon as Clarke was appointed. It was replaced by a formal government "advisory" committee consisting entirely of nongovernment members. This severed the link between the DOE fusion managers and the project directors that had been so effective in building consensus. The new advisory committee reported directly to Trivelpiece rather than to Clarke, further weakening the management structure put in place by Hirsch.

The new advisory committee was called the Magnetic Fusion Advisory Committee (MFAC) and was chaired by Ron Davidson. Davidson had been Director of the Applied Plasma Physics Division under Kintner (1976–1978) and was now head of the fusion effort at MIT. The first meeting of MFAC was held June 1–2, 1982, in Washington. During 1982, MFAC carried out four studies. Over the next several years, MFAC would carry out a series of (seemingly endless) technical studies, as requested by DOE. MFAC prepared 23 panel reports between 1982 and 1989. Trivelpiece and Clarke also commissioned the National Research Council (the operating arm of the National Academy of Sciences and the National Academy of Engineering) to host a 2-day workshop "on the identification of the future engineering needs for fusion." The workshop, chaired by Prof. Herb Woodson (University of Texas), was held August 3–4, 1982. The MFAC studies carried out from1982 to 1985 included:

- Evaluate several options for a possible upgrade of MFTF-B, 1983.
- Evaluate the long-term role of universities in the fusion program, 1983.
- Evaluate balance and priorities within the magnetic fusion program, 1983.
- Evaluate industrial participation in fusion energy development, 1984.
- Critically review DOE draft Magnetic Fusion Policy Plan, 1984.
- Assess scientific contributions from Alternate Concepts, 1984.

MFAC studies between 1985 and 1988 were published in the Journal of Fusion Energy [40].

On June 22, 1982, in Washington, an "Industry-Government Seminar on Fusion Energy Development" was held, jointly sponsored by Fusion Power Associates, the Atomic Industrial Forum, and the Electric Power Research Institute. DOE Director of Energy Research Al Trivelpiece told the group he had set up the MFAC to get responsibility, accountability, and perspective into the spectrum of unsolicited advice he was getting from a myriad of quarters. "A committee is not a substitute for effective management leadership," he said, "but it can certainly be a helpful tool." He said that because of the federal budget difficulties, "the idea that we may be able to proceed immediately with a Fusion Engineering Device is a little bit hard to imagine." He also warned that "some kind of decision with respect to tokamaks and mirrors may have to be made in the 1986–1987 timeframe."

Joe Gavin, President of Grumman Corporation, urged the group to plan aggressively. He said, "a shorter program will cost less than a longer program. If you plan it for 50 years, it will take 50 years. If you plan it for 10 years it might take 12 but at least it will be 12 instead of 50." Mel Gottlieb, Director of the Princeton fusion lab, told the group "You have to set yourself goals that are not easy." Having difficult-to-achieve goals "can easily make a difference of a factor of two in the rate of progress," Gottlieb said. Ron Davidson, the chair of MFAC, told the group "from a scientific viewpoint, the technical reasons for moving ahead promptly with an expanded engineering development program, as espoused by the recommendations of ERAB and the Fusion Engineering Act, are better founded today than 2 years ago."

Though the Reagan Administration was clearly not intending to implement the budget and schedule aspects of the Fusion Engineering Act, they nevertheless espoused the basic logic of transitioning to a more engineering-oriented program in the future, albeit on a delayed and less certain timetable. The House Science and Technology Committee continued to press the Administration to implement the fusion Act. Congressional hearings on fusion, at which I testified, were held on March 3, 1981, March 23, 1982, and October 20, 1983, for example.

On September 8, 1982, Sen. Pete V. Domenici (R-NM) and Rep. Marilyn Bouquard (D-TN) sponsored a Joint (House–Senate) Workshop on Fusion Energy. Donato Palumbo, head of the Euratom fusion program, and Seguri Mori, head of the Japanese program, were among those participating. John Clarke stated that although "the Administration recognizes that fusion is the principal energy option for the United States in the next Century and it recognizes the importance of this program," nevertheless "we must plan on relatively constant budgets for the FY 1984–1988 period." Clarke indicated that his strategy involved assuming a decision in 1988 to construct an engineering test reactor (ETR) and that Mirrors and Tokamaks would compete for the ETR, which would be built during the 1990s. This strategy represented approximately a 10-year slip from the 1976 plan and the Fusion Engineering Act.

On Christmas Eve 1982, the TFTR at Princeton, which began construction in 1976, made its first plasma. The MFTF-B mirror at Livermore was under construction, with completion still a few years away. The President's FY 1984 fusion budget request, submitted to Congress in February 1983, was for $467 million, compared to $461 million appropriated by Congress for FY 1983. The Congress would ultimately approve $469 million. Looking back historically in 2012, this would prove to be the peak DOE civilian fusion budget. Starting in FY 1985, the budget would go down every year for the next 7 years, reaching a (temporary) low of $284 million in FY 1991. The consequences for the logic and plans of the US fusion community were severe.

Fusion Power Associates held its third annual meeting and symposium on January 6–8, 1983. The theme was "Fusion 1983: A Symposium on the Readiness and Reasons for an Accelerated National Development Program." Leaders of the fusion program from throughout the USA came and shared their views and expertise. The first day consisted of review papers on all aspects of the fusion program, including TFTR, the Magnetic Mirror program (including MFTF-B construction), Bumpy Torus, Stellarator, Reversed Field Pinch, and Compact Fusion Concepts. A summary report was prepared [41] which includes keynote addresses by Herb Woodson and John Clarke, a luncheon address by Dr. Robert Sproull (President of the University of Rochester), and the complete prepared remarks from a panel discussion on "The Pace of Fusion Development." The panelists were Ron Davidson (MIT and MFAC chairman), John Emmett (head of the laser program at LLNL),

Ed Kintner, Mike McCormack, and Bob Hirsch. It was clear that optimism for rapid engineering development of fusion had begun to wane. Clarke indicated that the political reality was that the current fusion budget was judged to be fusion's "fair share" of the federal budget and, at that level, we could maintain a "base program" that would be "capable of making steady, demonstrable scientific and technological progress toward the ultimate goal" but that a "reactor engineering program" that included construction of a fusion engineering device or experimental test reactor and the attendant development program "would require a change in fusion's fair share of the federal budget." He said, "the option to move ahead aggressively with a reactor engineering program must be maintained" but that it is "only one option among many" and "moreover an option which may never be exercised, at least on a purely national basis." Others, like Emmett and Hirsch, emphasized that the fusion program needed to pay more attention to developing a reactor concept that would make fusion power "economically attractive."

In late 1982, Oak Ridge had conducted an internal review (headed by Lee Berry) of the data from EBT-S on which the design and approval of EBT-P had been based. This was further examined by a "data review panel" headed by Richard S. Post, who had a small mirror experiment at MIT. Both groups concluded that the EBT-S data had been misinterpreted previously and did not support the optimism on which EBT-P had been approved. During the week of February 15, 1983, DOE convened an EBT Senior Review Panel, chaired by Fred Ribe of Los Alamos. I served on the panel, along with Ron Davidson, Ken Fowler, and Herb Berk. We concurred with the previous two panels' analyses and recommended that DOE establish a set of milestones and performance goals for the EBT-S group. One year later, the EBT-P project was cancelled.

ERAB Review

In February 1983, the Secretary of Energy asked his Energy Research Advisory Board (ERAB) to conduct another fusion review (such ERAB reviews were mandated every 3 years by the Magnetic Fusion Energy Engineering Act of 1980). In his February 24 letter to ERAB requesting the review, then Secretary of Energy Donald Hodel acknowledged "the Department has been unable to implement fully some specifics of the Act due to budgetary limitations but has attempted to proceed, as much as possible, in accordance with the intent of the Act." He requested ERAB to give him their views on "the validity of the Department's approach, given the budgetary stringencies that make it impossible to implement all the requirements of the Act and the previous advice of ERAB." In addition to ERAB Chairman Lou Roddis, who would also chair the fusion review panel, the panel included perennial fusion reviewers Sol Buchsbaum and John Foster and also included Bill Stacey (Georgia Institute of Technology), a fusion researcher who had been active in the design of an international fusion Experimental Power Reactor called INTOR [30]. In a presentation to the ERAB panel August 30–31, 1983, I stated:

"Because of the rapidly evolving nature of proposals for next generation fusion devices, Fusion Power Associates has not taken a position on any particular new or upgraded devices. I remain convinced, however, that the principles I presented to the Buchsbaum (previous ERAB) panel 3 years ago remain valid today. These were:

- The purpose of the next major fusion device should be to extend the level of accomplishment beyond the maximum projected performance capability of TFTR, JET and JT-60 (the European and Japanese equivalents of TFTR).
- In addition to its detailed physics and engineering goals, the next major device should have one or more goals which can be easily perceived by the public as a step toward practical fusion power.
- The cost of the next device should not exceed approximately $1 billion.
- Important features of the next device, compared to its predecessors, should be improved reliability and availability, and the concept should have reasonable prospects of extrapolation toward a practical fusion power source.
- In August, also, the Fusion Power Associates Board of Directors issued the following Fusion Policy Statement:
- The development of fusion power should receive a higher national priority than it is now receiving. This is based on our belief that the availability and affordability of energy are essential elements to a healthy and vigorous industrial society and that an assured energy supply is an important determinant of world peace and security and a requirement for progress of future generations.
- The engineering aspects of fusion development should be given substantial additional effort now.
- Commitments should be made now to new and improved experimental facilities. This is based on our belief that such commitments are urgently needed to ensure continued program momentum and progress into the 1990s."

Fusion Power Associates institutional membership had grown from 13 charter members in late 1979 to 50 in early 1984, as shown in Table 5.3.

MFAC Strategy

In August 1983 also, the DOE's Magnetic Fusion Advisory Committee (MFAC), chaired by Ron Davidson of MIT, urged the DOE to adopt a "new strategy" having four principal features:

- Initiation in FY 1986 of a Tokamak Fusion Core Device (TFCD), a moderate cost device costing less than $1 billion, designed to achieve ignition and long pulse equilibrium burn.
- Potential utilization of the MFTF upgrade to provide a cost-effective means for quasi-steady-state testing of blanket and power-systems components, complementary to TFCD.
- Vigorous pursuit of the broad US base program in magnetic confinement.
- Utilization of the Development and Technology programs in plasma and magnet technology in support of specific hardware requirements for the TFCD and of other major fusion facilities, so as to minimize overall program costs.

Table 5.3 Fusion Power Associates Members, 1984

Applied Microwave Plasma Concepts	Kansas City Power and Light Company
ARCO Oil and Gas Company	Mathematical Sciences Northwest
BDM Corporation	Maxwell Laboratories, Inc.
Baltimore Gas and Electric Company	McDonnell Douglas Astronautics Company
Bechtel Group, Inc.	Northeast Utilities
Black and Veatch Consulting Engineers	Northern States Power Company
Boeing Engineering and Construction Company	Ontario Hydro
Boston Edison Company	Pennsylvania Power and Light Company
Burns and Roe, Inc.	Phillips Petroleum Company
Chevron Research Company	PSE&G Research Corporation
Combustion Engineering, Inc.	Science Applications, Inc.
EG&G Idaho, Inc.	Standard Oil Company of Ohio
Ebasco Services, Inc.	Stone and Webster Engineering Corporation
Electric Power Research Institute	Southern Company Services
Exxon Nuclear Company, Inc.	Texas Atomic Energy Research Foundation
Florida Power and Light Company	Tokamak Systems, Inc.
GA Technologies, Inc.	TRW, Inc.
General Dynamics Convair Division	United Engineers and Constructors
Gilbert/Commonwealth Engineers and Consultants	Universal Voltronics Corporation
Hipotronics	University of Illinois Nuclear Engineering Department
Hoya Optics, Inc.	University of Rochester Laboratory for Laser Energetics
ILC Technology	Varian Associates
Institut de Recherche d'Hydro-Quebec	W. J. Schafer Associates, Inc.
JAYCOR	Westinghouse Electric Corporation
KMS Fusion, Inc.	Wisconsin Power and Light Company

The MFAC stated that "carrying out the above program will require a budget increase of 25–40%, after inflation, over a period of several years." They recommended that a FY 1985 fusion budget of $535 million be sought. This latter recommendation was contrary to the flat FY 1985 budget guidance that had been provided to DOE by the OMB. In the end, the FY 1985 budget, as approved by Congress, was $430 million.

Ansel Adams

Famed nature photographer Ansel Adams took an interest in fusion in the early 1980s after visiting the LLNL and seeing the MFTF-B construction site (Fig. 5.1).

Adams later visited President Ronald Reagan and urged him to support the fusion effort. Adams joined Fusion Power Associates as an Individual Affiliate in mid-1983. In a letter to me dated July 18, 1983 (Fig. 5.2), Adams said, "I am sure you must know of my recent visit with President Reagan in Los Angeles.

Fig. 5.1 Photographer Ansel Adams on a visit to the fusion facilities at the Lawrence Livermore National Laboratory, talking to Ken Fowler, director of the fusion energy program there (photo by LLNL photographer James E. Stoots, Jr.)

I asked 'Why do you not take 10 or 20 billion dollars from your defense budget and apply it to a crash program for the development of fusion power? He gave no response to this although I added 'Its success would certainly free us from depending on imported fuels." Adams concluded, "I remain dedicated to the concept of fusion power (as a citizen) and I wish I was in a position to do more about it."

Adams scheduled a luncheon meeting with Fusion Power Associates Board of Directors in April 1984 near his home in California. He was forced by illness to cancel the meeting, and, sadly, he died 2 weeks later, on April 22, 1984, at the age of 82.

William R. (Bill) Ellis

In October 1983, Bill Ellis left his position as Director of the Magnetic Mirror Division in the fusion office at DOE to become Associate Director of Research at the US Naval Research Laboratory. Since 1975, when I hired him from the Los Alamos lab, Bill had been responsible for the evolution of the Magnetic Mirror program, including MFTF-B and, more recently, the Bumpy Torus program,

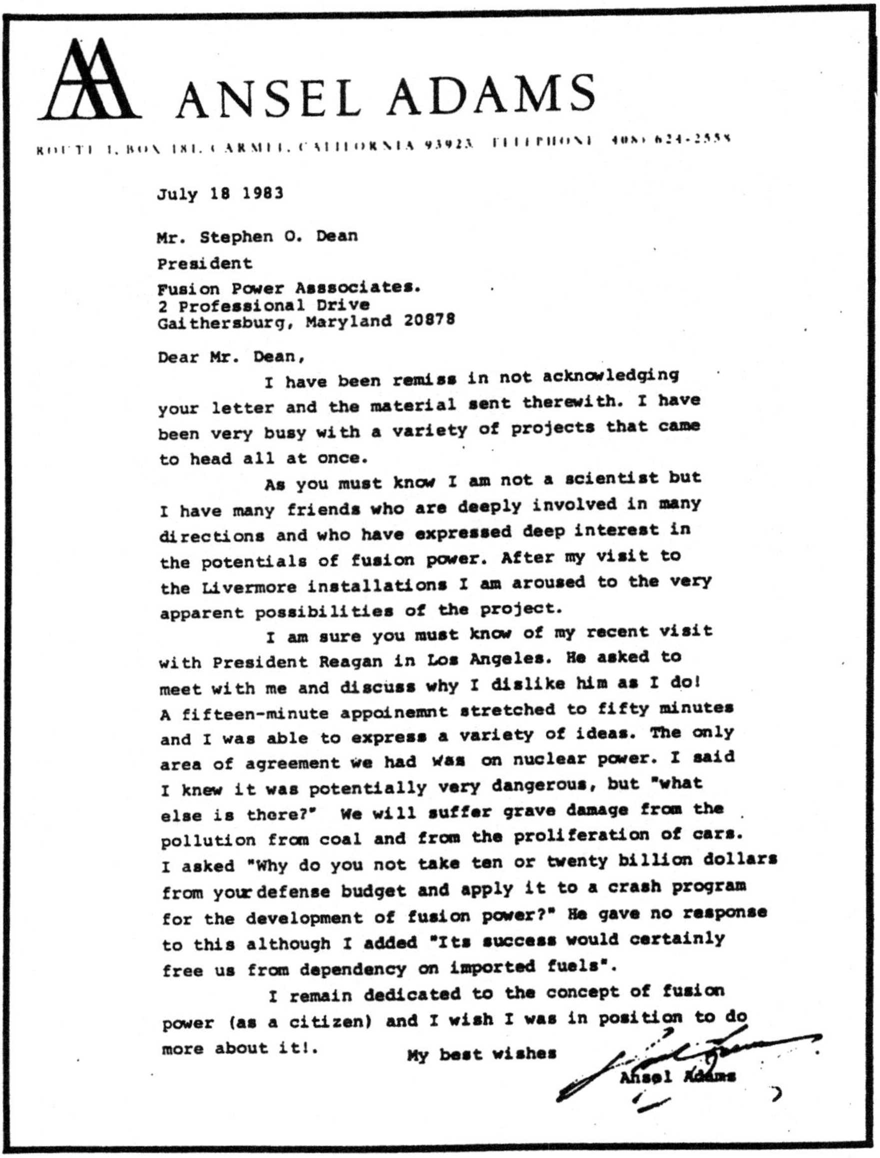

ANSEL ADAMS

ROUTE 1, BOX 181, CARMEL, CALIFORNIA 93923 TELEPHONE (408) 624-2558

July 18 1983

Mr. Stephen O. Dean
President
Fusion Power Asssociates.
2 Professional Drive
Gaithersburg, Maryland 20878

Dear Mr. Dean,

I have been remiss in not acknowledging your letter and the material sent therewith. I have been very busy with a variety of projects that came to head all at once.

As you must know I am not a scientist but I have many friends who are deeply involved in many directions and who have expressed deep interest in the potentials of fusion power. After my visit to the Livermore installations I am aroused to the very apparent possibilities of the project.

I am sure you must know of my recent visit with President Reagan in Los Angeles. He asked to meet with me and discuss why I dislike him as I do! A fifteen-minute appoinemnt stretched to fifty minutes and I was able to express a variety of ideas. The only area of agreement we had was on nuclear power. I said I knew it was potentially very dangerous, but "what else is there?" We will suffer grave damage from the pollution from coal and from the proliferation of cars. I asked "Why do you not take ten or twenty billion dollars from your defense budget and apply it to a crash program for the development of fusion power?" He gave no response to this although I added "Its success would certainly free us from dependency on imported fuels".

I remain dedicated to the concept of fusion power (as a citizen) and I wish I was in position to do more about it!.

My best wishes

Ansel Adams

Fig. 5.2 Famed nature photographer Ansel Adams became a fusion advocate, and he joined Fusion Power Associates as an Individual Affiliate in mid-1983

including EBT-P. The DOE awarded Ellis its Bronze Medal for Exceptional Service, "for his initiative and diplomacy which have led to worldwide recognition of key technical issues and have resulted in more efficient utilization of international mirror research efforts." In 1989, while at NRL, he received a

Presidential Senior Executive Service Award citing his "career achievements in the management and execution of research and development programs of national importance." He would later become Vice President of Ebasco Services, Inc., and chair of the US ITER Industry Council.

Major Progress Event: Lawson nτ Achieved at MIT

In late 1983, an event of major historical significance occurred when the Alcator tokamak group at MIT achieved the Lawson product of density and confinement time (nτ) of the plasma [42]. The Lawson product states that unless a certain value of this product is achieved or exceeded, net energy from fusion cannot be obtained regardless of how high a temperature is achieved. The group, headed by Ron Parker, included Martin Greenwald, Dave Gwinn, Steve Wolfe, and many supporting scientists, engineers, technicians, and supporting staff. The minimum "ideal ignition" temperature had been achieved previously in the PLT tokamak at Princeton in 1978. Thus, we had achieved in the tokamak both the temperatures and the Lawson product required for net fusion, but not simultaneously. There was no doubt in anyone's mind, however, that this could be achieved in a future device.

Tokamak Scaling

In the early 1980s, many small and mid-sized tokamaks were in operation in several countries. PPPL scientist Rob Goldston, who would later become PPPL director, surveyed all the experimental tokamak data and came up with an empirical scaling law for confinement time which fit existing data. It was then used for predicting plasma conditions in larger devices [43]. This became known as "Goldston Scaling" and has formed the basis for an evolving empirical design procedure, the latest of which, called ITER scaling, led to the current ITER design [44].

More on Inertial Confinement

Inertial confinement fusion (ICF) was funded within the DOE's weapons programs and was not part of the official civilian fusion energy program. The logic was that the experiments were designed to study the physics of the ignited fusion core of a hydrogen bomb. The ICF program was, consequently, primarily carried out at the DOE weapons laboratories of Lawrence Livermore, Los Alamos, and Sandia laboratories. The ICF program did, however, support smaller programs at the University of Rochester Laboratory for Laser Energetics and the US Naval Research Laboratory. There was also a privately funded effort at KMS Fusion in Michigan.

ICF was first recognized as a separate budget item in FY 1963, a few years after the invention of the laser. Since that time, it had grown slowly but systematically, reaching $209 million in FY 1982. A sequence of ever larger lasers and pulsed power facilities had been built. Beginning in FY 1984, however, the ICF budget began a 5-year decline to a (temporary) low of $155 million in FY 1987.

Since its inception in 1979, Fusion Power Associates had always advocated that DOE recognize and fund ICF research and development related to its energy potential. The scientists working on ICF were largely motivated by its energy potential, even while working on weapons physics and under DOE weapons funding. The University of Rochester and KMS Fusion were members of Fusion Power Associates.

The ICF budget had been ramped down from $209 million in FY 1982 to $170 million in FY 1984. When DOE submitted its FY 1985 budget in early 1984, it requested Congress to cut it further to $138 million. However, Congress had been adding monies to previous low ICF budget requests from the Administration for several years.

I was invited by the House Armed Services Committee to provide a statement on the Administration's FY 1985 ICF budget request. I provided the following statement, with identical statements going to the Senate Energy Committee and the House and Senate Appropriations Committees:

> "I am deeply concerned with the Department of Energy's FY 1985 budget request for inertial confinement fusion. For several years now the Department has systematically reduced their requests for operating funds in this program relative to prior years. Only the foresight and vision of the Congress, which has regularly added money to the DOE requests, has kept us from losing considerable numbers of talented scientists from this very productive program."
>
> "The people working in this program have constructed and operated laser and particle beam facilities at power and energy levels which seemed impossible 15 years ago. They have extended our understanding of the interactions of intense laser and particle beams with matter. In the process, important additional benefits have been spawned, including the new field of laser isotope separation, studies of weapons-related phenomena and direct energy beam weapons technology."
>
> "In light of the outstanding record of accomplishments established by workers in this field, it is incredible to me to witness the short-sighted budget cutbacks imposed on this program within a DOE defense budget which is expanding enormously." The FY 1985 DOE weapons budget requested a $1 billion increase over FY 1984
>
> "I urge the Congress to continue to provide the support necessary to maintain momentum, progress and enthusiasm within the inertial confinement fusion program."
>
> "I also urge the committee to maintain an interest in the civilian applications of inertial fusion, which I believe remain promising."

The Congress appropriated $168 million for ICF for FY 1985, compared to the $138 million DOE had requested and compared to the $170 million appropriated in FY 1984. The ICF budget would however decline to $155 million in FY 1986 and 1987 before beginning a gradual upward trend through FY 1993.

Industrial Participation

The question of the extent of industrial participation in fusion energy development has always been a difficult one. The AEC/ERDA/DOE culture has always favored its own laboratories and the universities. The laboratories themselves had extensive in-house engineering and component fabrication capability. The research divisions at AEC/ERDA/DOE, in particular, did not have a history of supporting work in industry.

On the other hand, if fusion was eventually to become a commercial reality, at some point knowledge and capability related to fusion technologies would have to be in industry. Also, industry was the link to the electrical utilities, presumably the ultimate customer.

We had built up a fusion capability at General Atomics, dating back to the mid-1960s. McDonnell Douglas was gaining a foothold, assuming the EBT-P project continued and was successful. Ebasco and Grumman were gaining experience in building TFTR. Other companies were building superconducting magnets for testing at Oak Ridge. Most other industries were either specializing in certain types of needed components for the laboratory or university fusion experiments or were anxiously awaiting the opportunities that construction of a fusion engineering test facility would provide.

In May 1984, the MFAC completed a study on "Industrial Participation in Fusion Energy Development." It stated in part, "It is not too early to begin to prepare industry for the role of prime contractor and major subsystem supplier to the government and to the utilities." They said, "There should be significant industrial participation in developing the fusion engineering database and in major projects such as TFCX (Tokamak Fusion Core Experiment, a proposed new facility) so that the private sector is better able to make decisions on the commercial potential of fusion energy." The MFAC said, "An engineering program should be maintained under all budget assumptions as a complement to other program elements."

Retrenchment

The ERAB fusion review panel issued its report [45] to the Secretary of Energy in February 1984, recommending that the department construct a fusion Burning Core Experiment (TFCX, but renamed BCX) "to explore ignition physics as a high priority in the fusion program." The report states, "The BCX should specifically provide for future upgrades of the facility and, in addition, should incorporate, as far as practicable, those component features that can serve as a focus for future reactor technologies." They also said, "The Board strongly urges that the United States vigorously seek bilateral or multilateral collaboration for the development, construction and operation of the proposed BPX. Such collaboration would permit more aggressive technological objectives and accelerate the time frame for proving the technological feasibility of fusion for the benefit of mankind." The ERAB seemed to envisage the BCX as a facility that could evolve into a facility more like the engineering test reactor in the Fusion Engineering Act, whereas the TFCX the US fusion community was advocating at the time had more limited physics goals.

Though the first few years (1981–1983) of the Reagan Administration did not provide the fusion program budget growth and engineering focus envisaged in the Fusion Engineering Act; nevertheless, the budget had grown slightly, and the program was vigorously planning for construction of a new "burning plasma" facility to outperform TFTR. In mid-1984, however, the Congress cut the FY 1985 fusion budget to $430 million from its FY 1984 level of $470 million. The DOE absorbed this cut by delaying purchase of the tritium system for TFTR by 2 years, by further slowing the construction of MFTF-B, and by making cuts in other parts of the fusion effort. The Administration also began to tell the fusion community to abandon its engineering and goal-oriented planning focus. Speaking to an MFAC meeting on behalf of the President's Science Advisor, George Keyworth, Ralph DeVries (OSTP Assistant Director for General Science) suggested a new goal statement for the fusion program: "A well-balanced program that combines the scientific richness and inherent relevance of plasma physics, and the production of well-qualified talent, with the prospect of achieving a unique energy resource in the future." He said that only the Applied Plasma Physics portion of the fusion budget was described in a way that its objectives "are clearly aligned with the Administration policy." He particularly indicated that "large reactor-like machines" were not going to be supported by the Administration.

After assessing the budget and policy situation at its July 18, 1984, meeting, the MFAC concluded that the prospects for proceeding with a TFCX/BCX size facility had "dimmed dramatically, partly as a result of the FY 1985 budget cut and partly because of the Executive Branch attitude on large fusion experiments." MFAC was not ready to completely abandon the need for new facilities, however. They asked the community to develop two new options: "(1) a copper-coil design that is optimized for minimum cost, while meeting the basic scientific objectives (of TFCS/BCX) and (2) a superconducting-coil design that is optimized for the highest possible scientific and technological relevance and upgradability." At that MFAC meeting, Bruno Coppi of MIT suggested the "Ignitor" copper-coil ignition tokamak. Speaking in the public comment portion of the meeting, I urged the DOE to establish a Senior Review Panel to review Coppi's ideas and to report its findings "within 6 weeks, not 6 months." I expressed "continued dismay at the US government's lackadaisical plans for the future and its pre-occupation with the seemingly endless series of design iterations and committee studies as a substitute for making decisions, building experiments and getting results." "We are spending a million dollars a day while this process goes on," I said [46].

Trivelpiece Interview

Al Trivelpiece, the Director of the Office of Energy Research at DOE, offered to provide me with an exclusive interview after the Congressional actions on the FY 1985 budget. That interview, published in the September 1984 issue of Fusion Power Associates Executive Newsletter [47], is provided in its entirety below:

> Congress recently passed an FY 1985 budget for magnetic fusion that was $43 million less than the President requested. Fusion Power Associates president Steve Dean discussed the DOE's reactions to the cuts and their vision of the future with DOE Director of Energy Research Alvin W. Trivelpiece (AWT).

Dean: Were you surprised when Congress cut the magnetic fusion budget and what do you think was the cause?

AWT: I was both surprised and disappointed. For some time the fusion program has been funded at less than cost of living increases. The last few years this has been due in part to overall budget stringencies for federal energy programs. The Congress has generally added funds to the Administration's request for fusion research; however, I was pleased that for FY 1985 the Administration asked for a solid program budget of $483M. The House authorization committee even favored increasing the President's budget by $22M. However, the House appropriations committee didn't see it that way. They recommended a $64M cut. The Senate proposed only a $10M cut, but in conference they compromised with a $43M cut which Congress then passed, leading to a $440M FY 1985 budget. There are many versions of why this happened. It is not productive to speculate which version may be right. However, it is my opinion that several members of Congress felt that the fusion program had not adjusted to the realities of the present energy supply situation and the budget deficits. I was told they "wanted to get our attention." They certainly got it!

Dean: How are you going to accommodate the budget cut?

AWT: After a careful review of where the program is and how it got there, I decided that the best course of action was to take the cut in three places, but primarily from two large projects: TFTR and MFTF-B. First, we will delay the D–T (deuterium–tritium) burning experiment previously scheduled in TFTR for 1986. This will save program funds in two ways. One, it eliminates certain tritium and remote handling costs that would have to be incurred before doing the experiments, and two, it eliminates the need to have provided additional funds in FY 1985 that would have been needed to maintain the 1986 D–T goal. The actual cut in the TFTR funding will be $9M. This delay in D–T burning of a few years will permit critical hydrogen experiments to be done with much greater efficiency and less inconvenience than would be the case after tritium is used since we would then require remote handling. I want to stress that this is merely a delay.

Conducting an energy breakeven experiment utilizing DT in the TFTR is still a critical near-term milestone for the U. S. fusion program.

Second, I decided to reduce the MFTF-B operating budget by $15M. This will delay some of the FY88 goals to 1989 or 1990. Here again, scientific objectives remain intact; however, we are extending the timeframe for their accomplishment. We have expended a great deal of time, effort and fiscal resources to define and fabricate the hardware necessary to develop the physics of linear and toroidal (tokamak) systems. It would be folly not to exploit the physics capabilities of these devices. As with any program of experimental research, future directions for these systems will depend upon the nature and significance of the scientific results achieved.

Third, I decided to reduce by $13M those development and technology activities that were not directed to support of the nearer-term aspects of the program. I recognize that we must do systems studies to give us perspective. I don't like taking cuts in the technology programs because it is simplistic to think that all good science in fusion is being done only by plasma physicists. Much of our technology development is outstanding "science" in the broad meaning of the term.

Those were difficult choices to make and were not taken lightly. Given the present circumstance, I believe these actions are the best way to preserve the essential course of the program in a reduced budget situation.

Dean: Are we likely to get cut further in FY 1986?

AWT: I hope not. The Department is putting together its FY 1986 submission to OMB now and many things can happen between now and when the President submits his budget to Congress. I believe the program can justify the money it is getting in terms of the outstanding scientific progress and the results it has achieved. Fusion is excellent science, forefront technology and is making steady progress.

Dean: I've heard a concern expressed in the (fusion) community that the fusion program may lose its focus and become a "science-only" program. What is your view on that?

AWT: The fusion program is a mission-oriented energy program and must remain so. Its aim is to ultimately provide the world with an environmentally attractive energy source. This is its long-term goal. However, the need to achieve that goal is some time in the future, and in the near-term, we have a program that is at the forefront of many areas of advanced science and technology. We intend to emphasize the value of the research. The fact that the fusion program is an excellent research effort has gotten somewhat lost in all the energy rhetoric.

In terms of funding, the key term is "balance." I realize that much of what we do in the technology area is either absolutely essential to the scientific experiments or in many cases is contributing to our progress as a high technology society.

Dean: What about university programs? Are they more important than laboratory or industry programs?

AWT: Again I want to emphasize our intent to maintain "balance." The fusion program has been one of the outstanding examples of a federal program that consistently provided universities and students opportunities and funding to do forefront research. Since 1965, over 1,100 students have received Ph.D.'s as a result of fusion support at universities. Fusion also supports graduate fellowships in our engineering schools. About half of these 1,100 students work in high tech industries that contribute to our national competitiveness and security.

Dean: What about industry? Is there a role for industry in the near-term?

AWT: Industry has played and will continue to play many varied roles, from component supplier, to R&D, to manufacturer. It's probably true that without a near-term large construction project the dollar value and visibility of industrial opportunities will be reduced. However, this does not mean that industry shouldn't be involved in the program. Quite the contrary, the involvement of industry is essential to the fusion program. I hope the fusion program will retain strong industrial participation, but the budgetary pressures and near-term priorities will slow this up, along with the slowing up that will occur in other parts of the program.

Dean: Do any of your recent actions imply that you are discouraged by the prospects of the tokamak concept and feel we need to find a better alternative?

AWT: We must continually look for better ideas and I intend to expand our support of other concepts to develop the most promising approach to fusion. At the same time, there is plenty of evidence that the tokamak provides just a good a starting point for improvement and innovation as other concepts. The tokamak provides a standard and a challenge for those who advocate other concepts.

Dean: You just came back from an economic summit follow-on meeting on fusion cooperation in Brussels. What happened at that meeting and does this mean an international fusion project is possible?

AWT: Since the Versaille summit, the economic summit process has focused some attention on science and technology. The Versaille summit defined 18 areas of science and technology for which international cooperation or collaboration would be appropriate. Fusion is one. We have the co-lead in this area along with the Commission of the European Communities. We met at Washington, DC, last November following the Williamsburg summit and in anticipation of the London summit. Based on the guidance from the June London summit, we met in Brussels in July. What we did at the Brussels summit was to establish three subcommittees. One is to be concerned with collaboration on major future facilities. The other two will deal with administrative and technical problems that could impede cooperation. These subcommittees are to report back to the parent committee by January as part of the preparation for the 1985 economic summit in Bonn. This summit process has created a new channel of communication and involvement at higher political levels than has existed in the past. At higher political levels there is a general concern that there not be duplication of facilities that cost in excess of $1 billion.

I believe that we need a plan that outlines steps required to make fusion work without regard to when and where the work is done. The agreement needs to be developed to do the work world-wide in such a way that duplication of effort is avoided. All of this is difficult and time-consuming. The first few steps using the summit process have been taken. This could lead to a program of international collaboration where greater progress can be made without enlarged in-country levels of support. This may take the form of several bilateral or multi-lateral programs similar to the one the Japanese have with us on the Doublet program at GA Technologies.

Dean: What about the future?

AWT: I am confident that fusion is scientifically and technically possible. I believe that it will be an important future energy option. In the near-term we need to continue to pursue research to uncover the best ideas. In the long-term it has to compete economically with other energy options. The energy crisis of 1975 sent us off on a path that would have had fusion reactors operating by the year 2000. But to do that requires a commitment of resources that are not likely to be forthcoming in the present energy and budgetary climate. A revised plan for the program that takes this reality into account and involves international collaboration needs to be developed. Such a planning effort is underway.

Budgets and Reality

By the mid-1980s, it was clear, as is evident in Fig. 5.3, that the fusion budget was not on a trajectory to result in a fusion power plant anywhere near the year 2000. In mid-1984, the fusion program was told to stop designing the TFCX and to refocus on less expensive facility options and international collaborations. In early 1985, the President requested FY 1986 fusion funding of only $390M compared to the $430M appropriated by Congress for FY 1985. The Congress would eventually reduce that to $362M (as part of Newt Gingrich's "Contract with America"). The FY 1984 budget had been $469M. Thus, in a 2-year span, the fusion budget was reduced by about $100M. This created a crisis in mid-1985. The separately budgeted inertial confinement fusion effort likewise was being reduced from its high of $209M in FY 1982 to $155M in FY 1986.

Although the magnetic fusion community had lost hope that it would be allowed to construct an engineering test reactor, or even a TFCX, it was still being encouraged to design and propose a less expensive tokamak device that would outperform the TFTR. TFTR, at that point, had had its DT fusion phase delayed until 1989 (it would later be delayed further). By late 1984, as reported at the IAEA biannual fusion conference in London, significant progress was being made across the board: both TFTR and JET were achieving record plasma confinement time in hydrogen plasmas and TFTR neutral beam heating had resulted in plasma temperatures above

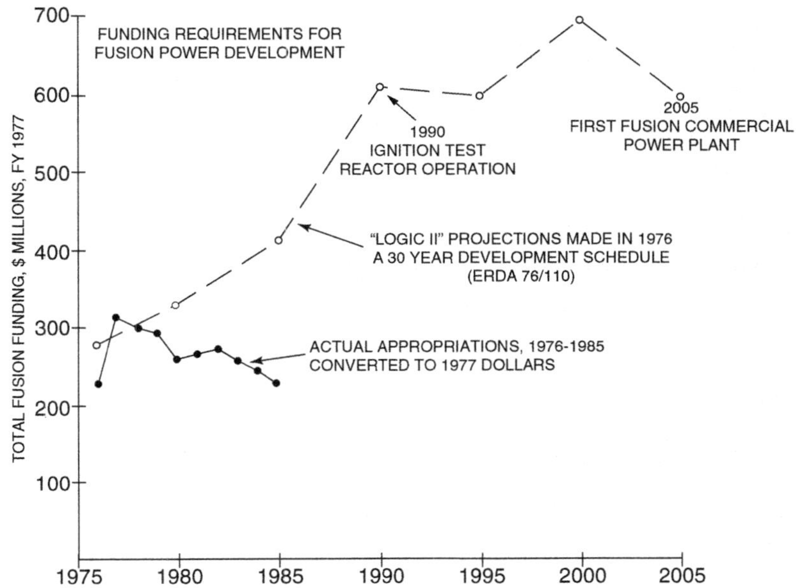

Fig. 5.3 Fusion budget history, 1975–1985. *Upper curve* shows the funding required under the 1976 plan for "Logic II" (operating demonstration power plant in 2005). *Lower curve* show actual funding provided 1976–1985 (Fusion Power Associates Executive Newsletter, Nov 1984)

3 keV (and would further raise this to 17 keV by July 1986); the Doublet III tokamak had achieved a high value of the triple product of temperature, density and confinement time ($Tn\tau$) of 4×10^{13} keV cm^{-3} s (compared to a value of about 10^{15} needed for a fusion power plant); the Alcator-C tokamak had achieved the Lawson $n\tau$ parameter of 10^{14} cm^{-3} s; and in the inertial confinement fusion program, the Novette laser had compressed a fusion capsule to 140 times liquid density (compared to about 1,000 times liquid required for a fusion power plant). From a technical standpoint, the fusion program was ready for acceleration towards its practical goal. Such hopes were not to be realized, however.

Successes and Disasters: 1985–1989 — 6

It's perfectly natural to want all the facts and to hold out for the research that guarantees a particular program will work. That's fine in theory, but real life just doesn't work that way. At some point, you've got to take that leap of faith.

–Lee Iacocca
Chairman, Chrysler Corporation

In the FY 1986 budget submission to Congress in early 1985, President Reagan proposed to cut the magnetic fusion budget by another $50 M, to $390 M (recall that the budget had been reduced already by about $40 M in FY 1985 compared to FY 1984). He also proposed to cut the inertial confinement fusion program more than in half (from $168 M to $70 M). By the time Congress got through with this submission, the magnetic fusion FY 1986 budget had been cut even further, to $362 M, but the proposed inertial fusion cut had been largely restored (to $155 M). DOE, in its inertial fusion budget submission, had proposed to eliminate inertial confinement fusion as a separate program altogether and simply "bury" it within its multibillion-dollar weapons R&D category. Congress refused to go along with that proposal.

On March 5, 1985, I gave the luncheon address to the American Nuclear Society Sixth Topical Meeting on the Technology of Fusion in San Francisco. The title of my talk was "What Will It Take to Accelerate the Fusion Program." I said, in part:

> "I am frustrated, as a lot of you are, that the fusion program is not able to move the way we all know it could move and the way it should move. My message is that I think there are three necessary and sufficient conditions for accelerating the fusion program."
>
> "The first necessary condition is that there has to be *systematic progress*. We have to show, and we have been able to show, that we are moving along the path to fusion power."
>
> "The second necessary condition is that we must have conceptual designs of fusion reactors that are acknowledged to look attractive as potential commercial products. We must have an embodiment, or several embodiments, that we are proud of, of which we could say, 'If we can do that, somebody would want it.' I wish I could say that we had such a design in the same way I can say we've had systematic progress."

S.O. Dean, *Search for the Ultimate Energy Source: A History of the U.S. Fusion Energy Program*, Green Energy and Technology, DOI 10.1007/978-1-4614-6037-4_6,

"The third ingredient is that there has to be in the country a level of total R&D on energy technology that is commensurate with the energy problem that the country's going to face down the road. Fusion is not going to get accelerated because it is fusion. Fusion prospered in the 1970s, and will prosper again, when the country realizes that energy technology has to be carried out at a level of financing and effort that is commensurate with the fact that it is going to impact an industry that is multi-trillions of dollars of investment in the 21st century."

Regrouping

Several versions of a high-magnetic-field "small" ignition tokamak were proposed, based on copper-coil (i.e., nonsuperconducting) magnets. These were similar in design to the Alcator experiment at MIT but varied in size, magnet strength, etc. Bruno Coppi and Dan Cohn (MIT), Y-K Martin Peng (ORNL), and scientists at Princeton were among those proposing such designs. The international community, via the IAEA-sponsored INTOR collaboration [30] that began in 1978, was also suggesting a large, superconducting, ignited-plasma engineering test reactor to be built as an international venture among several countries. Mike Roberts, in the DOE Office of Fusion Energy, was especially effective in encouraging the latter and received the DOE's Exceptional Service Award in early 1985 for his efforts. Meanwhile, France had constructed a midsized superconducting tokamak, called Tore Supra, and China was operating its first tokamak.

An op-ed by Flora Lewis in the April 12, 1985 New York Times reported on a "Consultation on International Security and Arms Control" cohosted by former US Presidents Gerald Ford and Jimmy Carter at Carter's center at Emory University in Atlanta. She reported that participants (which included Evgeny Velikhov, vice president of the USSR Academy of Sciences), looking "to thaw what Mr. Gorbachev has called the recent ice age in Soviet-American relations," felt that a joint program for fusion energy development would be "an excellent candidate for an exciting, constructive accord" that could be agreed to at the Gorbachev–Reagan summit that both countries hoped would occur in the fall. Lewis opined, "putting fusion on the (summit) agenda won't avoid all the abrasive issues, but it would promise at least one important success and brighten the horizon for the tougher talks." Her comment would prove to be prophetic.

In June 1985, a panel of senior tokamak scientists and engineers from the USA, Canada, Europe, and Japan, chaired by Ron Davidson (MIT), conducted an "Ignition Design Point Workshop." They concluded [48] that "the present data base is sufficient to guide the conceptual design of a tokamak ignition experiment" and that "the ignition experiment could provide unique and highly useful information." They did not express a preference among the three design options reviewed but recommended proceeding with "conceptual design of a compact ignition device."

In the summer of 1985, the DOE Office of Fusion Energy established a Technical Planning Activity (TPA) under the direction of Dr. Charles Baker (Argonne National Laboratory). The group was charged to identify program elements and subelements to resolve technical issues; prepare statements of objectives for the program elements and subelements; develop technical plans for each program element, including identification of necessary facilities; identify resource requirements; and stimulate community consensus and recommendations for priorities within program elements.

Under Baker's overall leadership, the work was organized into three groups: plasma science (headed by Jim Callen, U. Wisconsin), fusion technology (headed by Mohamed Abdou, UCLA), and fusion systems (which I headed). The study was completed in January 1987 [49]. The fusion systems' portion was published in J. of Fusion Energy [50]. Baker, Callen, Abdou, and I received Department of Energy Distinguished Associate Awards in March 1988 for our efforts.

Reagan–Gorbachev Summit Meeting

The Reagan–Gorbachev summit meeting was held in November 1985 and produced a landmark agreement regarding fusion. The official joint statement said, "The two leaders emphasized the potential importance of the work aimed at utilizing controlled thermonuclear fusion for peaceful purposes and, in this connection, advocated the widest practicable development of international cooperation in obtaining this source of energy, which is essentially inexhaustible, for the benefit of all mankind." In a nationally televised report to a joint session of Congress, Reagan said [51], "As a potential way of dealing with the energy needs of the world of the future, we have also advocated international cooperation to explore the feasibility of developing fusion energy." A US State Department official involved in follow-up planning to the summit meeting told Fusion Power Associates [52] that Reagan had stated, "I want to build a fusion reactor."

President Reagan, however, neglected to tell his Office of Management and Budget that such an international fusion effort would cost money. When the President's FY 1987 budget went to Congress in early 1986, it proposed an additional \$50 M cut to the magnetic fusion program (down to \$333 M) and once again to decimate the inertial confinement fusion program (taking it down from \$155 M to just \$22 M) and also once again proposed to "bury" its remains within the DOE's multibillion-dollar weapons budget (Congress again refused the latter). The budget document stated that the US fusion magnetic mirror program would be terminated.

The Reagan–Gorbachev summit meeting agreement on fusion kicked off a series of international discussions at high levels. Fusion became the subject of a summit working group under the aegis of economic summit meetings that occurred during the first half of 1986. Al Trivelpiece served as the US representative on this group. The idea of jointly constructing a fusion engineering test reactor was taking hold among this group. However, the concept of jointly constructing a fusion engineering reactor with the Soviets was running into opposition at the Defense Department as part of the Administration's internal coordination of policy among agencies. Pentagon staff reporting to Defense Department Assistant Secretary for International Security Policy Richard N. Perle actively sought through the National Security Council to stop President Reagan from concluding an agreement with the USSR to construct a fusion reactor [53]. Their objections were based on opposition to "technology transfer" of US know-how to the Soviets that might provide the Soviets with skills and technology that would threaten the security of the United States. Energy Secretary John Herrington sided with the Defense Department. This effectively transformed the fusion cooperation from one which could have built the facility during the 1990s (10 years later than in the Magnetic Fusion Energy Engineering Act of 1980) to a design-only collaboration. The design, as it turned out, would go on for most of the next 20 years, before construction finally began.

The Declaration of Energy Independence

Various organizations with energy interests, including Fusion Power Associates, had been pressing the Reagan Administration to take a more active interest in energy matters. The organizations jointly prepared a "Declaration of Energy Independence" and the then Vice President George H. W. Bush signed it in his White House West Wing conference room (Fig. 6.1).

About this time (1985–1986), the Reagan Administration also "deregulated" the US electric power industry with the result that utility managers became very short-term profit oriented. Previously, local and state regulators had ensured that long-term payoff research and development funds were included in the customer rate base. Most electric utilities then abolished their internal R&D departments and the utility-sponsored Electric Power Research Institute (EPRI) canceled its small fusion research effort.

Fig. 6.1 The author looks on as the then Vice President George H. W. Bush signs the "Declaration of Energy Independence" prepared by a group of energy organizations, including Fusion Power Associates

Major Achievement: TFTR

During 1985–1986, scientists at the Princeton Plasma Physics Laboratory, led by Don Grove and Dale Meade, achieved many of the original deuterium plasma objectives of TFTR. The machine was operating at its design specifications of magnetic field and current; the plasma density–confinement time product had exceeded the Lawson value previously obtained in the high-field Alcator device at MIT; and plasma temperatures exceeding the design goal of 10 keV were obtained. Figure 6.2 shows the interior of the TFTR vacuum chamber. The experiments carried out with deuterium fuel had an equivalent DT gain of 0.2.

Fig. 6.2 A look inside the TFTR vacuum vessel

Inertial Confinement Fusion Review

In February 1985, at the request of Congress, the White House Office of Science and Technology Policy (OSTP) initiated a review of the DOE inertial confinement fusion (ICF) program, "both the classified and unclassified aspects." At the time, the DOE and OMB were seeking to make drastic cuts in the ICF budget and to "bury" it within the multibillion-dollar DOE weapons budget. The OSTP assigned the review to the National Academy of Sciences, via a panel chaired by Prof. William Happer of Princeton University. The Congress hoped to buttress its own stand against the DOE/OMB actions via the academy review, while the Administration (OSTP) hoped for just the opposite. The academy was asked to provide its report by October 1985.

In August 1985, the panel provided an interim report to OSTP that strongly praised the technical accomplishments, progress, and prospects for success of the ICF program. The OSTP was not happy with the positive tone of the report since it did not buttress the Administration's plan to cut and bury the ICF program in the DOE weapons budget. The OSTP refused to make the interim report available to the public. The DOE Office of Classification also objected to statements in the interim report stating that ICF is overclassified to the detriment of scientific progress [54]. A member of the panel told Fusion Power Associates [55], "The Committee has not and will not change a word in its interim report." The panel member also said, "We listened to the Office of Classification for two hours and, when the session ended, we were more convinced than ever that we were right in criticizing their policies."

As 1985 came to a close, neither the interim nor a final academy report on ICF had surfaced. In December 1985, I filed a Freedom of Information request with OSTP asking for the interim report. In a letter dated February 10, 1986, I received a response from Jerry D. Jennings, executive director of OSTP, stating, "At the present time we are denying the release of the interim report of the National Research Council's panel to review the DOE's inertial confinement fusion program because the report contains information that requires coordination with other agencies. We will notify you as soon as we receive the results of the coordination and of our decision concerning release at that time."

The Congress was not happy with this situation. They requested that the ICF review committee chairman, William Happer, appear at a hearing. The executive branch initially refused to permit Dr. Happer to testify but relented under threat of congressional subpoena. What follows are verbatim excerpts from Prof. Happer's testimony "On the Interim Report of the National Academy of Sciences' Committee for a Review of the Department of Energy's Inertial Confinement Fusion Program":

> "Certainly we were persuaded that the ICF Program is today a vigorous and successful research effort which has made striking progress over the past 5 years, and that the outlook for success is more optimistic today than at the time of the last review (1981)."
>
> "The Committee was very favorably impressed by the quality of research facilities and the work being carried out. It was particularly impressed by the caliber and motivation of the research teams assembled at each of the research centers. These teams are a national resource not easily or quickly reassembled once they are disbanded."
>
> "If the decision point referenced earlier is to be reached in a timely manner (5 years), steady, rational funding over the next few years is essential. The ICF Program has traditionally been identified as a line item in the DOE budget. The majority of the Committee feels that this program identity should be maintained. There are serious problems with including the ICF Program in the RD&E portion of the DOE weapons program."
>
> "Separate line-item funding of the ICF Program would . . . help DOE evaluate and guide the program through the next five years . . ."
>
> "The only consistent external criticism of our interim report has been that it is too optimistic and that our credibility will suffer because of it. Since we do not wish to waste the efforts of a group of very busy and very dedicated committee members, nor the work of scores of people who supported our review, we redoubled our effort to find technical flaws in the program. The final report will not be without criticism. However, that redoubled effort produced nothing that would alter my testimony here today."

Major Achievement: Nova

In the spring of 1986, scientists at the Lawrence Livermore National Laboratory, using the new Nova laser, achieved the Lawson parameter, $n\tau$, reaching a value of 2×10^{14} cm^{-3} s. The temperature achieved was 1.5 keV, only a factor of about 3 below the Lawson minimum ignition temperature of 5 keV. The Lawson $n\tau$ previously had been achieved in MIT's Alcator tokamak and in TFTR.

In April 1986, the OSTP released the ICF committee's interim report to the public. In addition to the points covered in Happer's congressional testimony, the interim report stated: "Classification of much of the ICF Program is a difficult problem which is hindering progress by restricting the flow of information. These restrictions are hurting the morale of imaginative scientists . . . . Classification also keeps the scientific public from fully appreciating the important progress which has been made by the ICF Program, or from criticizing its weaker parts. A more widespread understanding of the achievements of ICF would make it easier to support the higher-priority parts and to de-emphasize the less urgent parts of the program. We recommend the formation of a high-level committee to review the issue of ICF classification and to formulate new, more realistic and flexible classification guidelines." The OSTP also released the final report which, in all substantive respects, was identical to the interim report.

I had written a letter to DOE asking for major ICF declassification. I received a letter in response from Dr. Charles F. Gilbert, acting director, DOE Office of Classification, stating, "With respect to your comments on Inertial Confinement Fusion (ICF) classification policy, I share your concern that we may not be realistic in some areas of that program. To that end, we have begun this month (April 1986) a review of the Department of Energy ICF classification policy."

I also wrote a letter dated April 7 to Energy Secretary Herrington asking in part "that you insist that the total budget for inertial confinement fusion remain separately identified in the DOE budget and that your staff be held accountable for the progress of this program and for assessing its potential." In response, I received a letter dated April 17 from Major General George K. Withers (head of the DOE weapons program office) stating in part [56], "We do not agree with the National Academy of Sciences committee's recommendation for a separate line-item providing for inertial confinement fusion. Inertial confinement fusion must be viewed as an integral part of the weapons research, development and testing account with funding levels established based on technological benefits and overall research and development requirements and priorities."

Congress thought otherwise, however. Instead of providing ICF with the $24 M requested by the Administration, Congress provided $155 M. In doing so, the House Armed Services Committee in its report (99–718) stated, "The committee is concerned that the DOE budget request for fiscal year 1987 once again proposes to disestablish the Inertial Confinement Fusion (ICF) program. Under the DOE approach, ICF would cease to exist as an entity, its programmatic goals abandoned, its past accomplishments consigned to oblivion and its promising future left in doubt. The ICF program has been authorized by Congress since 1977 because it has required central dedicated management and research teams and adequate annual funding. To merge the program's management and funding at this point with that of the overall weapons research and development programs, as proposed by the Department of Energy, would consign the ICF program to administrative extinction. The committee cannot agree with this approach. The committee recommends that Congress continue to authorize appropriations for the ICF program as a separate line item."

Although in January 1987 the DOE would again recommend a reduction ($25 M) in the ICF budget (Congress would later add $41 M to the DOE's request), DOE would never again seek to suppress identification of ICF funding as a line item in the budget.

Halite–Centurion

Although a DOE program called Halite–Centurion had been described as important in several reviews of inertial confinement fusion during the 1980s, neither its nature nor its results had ever been described. This was because the program was (and remains) classified. DOE, however, was forced to acknowledge its existence and importance to ICF following a front-page story by William Broad in the New York Times, March 21, 1988. The article, headlined "Secret Advance in Nuclear Fusion Spurs a Dispute Among Scientists," states in part, "In top-secret experiments, Federal researchers have achieved one of the nation's most costly and elusive scientific goals: to ignite a nuclear fusion reaction in tiny pellets of hydrogen, producing powerful bursts of energy." The article states that "The success was achieved in unorthodox experiments some 2 years ago at the Government's underground nuclear test site in the Nevada desert, according to Federal scientists and officials, who spoke on the condition of anonymity." The article states, "fusion in tiny fuel pellets was triggered by a blast of radiation from an exploding nuclear weapon." The article claims that "The results have triggered a bitter dispute over how the field of small-scale fusion should progress." The article states that the joint LLNL–LANL program had been underway for a decade.

The Compact Ignition Tokamak

During 1985, several scientists within the magnetic fusion program proposed designs for "small" ignited tokamaks. In February 1986, the DOE's Magnetic Fusion Advisory Committee (MFAC) recommended proceeding with the Compact Ignition Tokamak (CIT). Acting on MFAC's advice, Al Trivelpiece and John Clarke at DOE endorsed the CIT, and it was proposed in the President's FY 1988 budget, submitted to Congress in January 1987. Because of the TFTR infrastructure (e.g., tritium handling equipment and power supplies) at the Princeton Plasma Physics Laboratory (PPPL), they were given responsibility for the project. John Schmidt, reporting to Don Grove, was named project manager. Trivelpiece, however, then left DOE in spring 1987 to become executive director of the American Association for the Advancement of Science (1 year later he became director of the Oak Ridge National Laboratory). His deputy, Jim Decker, took over in an acting capacity until Trivelpiece's replacement as director of DOE's Office of Energy Research, Robert Hunter, took office in the fall of 1988. Hunter was president and founder of Western Research Corporation, San Diego. He was also the brother of Congressman Duncan Hunter (R-CA). Hunter's appointment to replace Trivelpiece at DOE would later spell the death knell for CIT.

The International Thermonuclear Experimental Reactor

Following the Reagan–Gorbachev 1985 summit agreement to collaborate on developing a fusion engineering test reactor, international discussions during 1988 led to the inclusion of Europe and Japan and formation of a study center located at Garching, Germany. The device to be designed was named International Thermonuclear Experimental Reactor (ITER). Ken Tomabechi, from Japan, was named to head the study, assisted by John Gilleland (USA), Romano Toschi (Italy), and Y. A. Sokolov (USSR). Following another Reagan–Gorbachev summit meeting, December 7–10, 1988, in Washington, the two issued a joint summit statement that included the following: "They affirmed the intention of the US and the USSR to cooperate with the European Atomic Energy Community (EURATOM) and Japan, under the auspices of the International Atomic Energy Agency, in the quadripartite conceptual design of a fusion test reactor."

An international design team was established at Garching, Germany. The effort, called Conceptual Design Activities (CDA), began in 1987 and finished in 1990. The ITER collaboration built on the very successful INTOR international design collaboration that had been operational since 1978.

Musical Chairs

A number of important personnel changes occurred within the fusion program during the late 1980s. Following the "mothballing" of the MFTF-B and termination of the LLNL magnetic mirror program, LLNL fusion program director Ken Fowler would first be asked to head up a study to design a US-based engineering test reactor, and later, he would leave LLNL to head the Nuclear Engineering Department at the University of California at Berkeley; Richard Briggs succeeded Fowler, becoming LLNL Associate Director for Beam Research and Magnetic Fusion Energy; Marshall Rosenbluth would leave his post as head of the Fusion Research Institute at the

University of Texas to join the University of California at San Diego and General Atomics; Fowler's mirror program deputy at LLNL, David Baldwin, left LLNL to replace Rosenbluth at the University of Texas (where they would propose Ignitex, an alternative design to the CIT); Tom Simonen, head of the terminated TMX-U mirror experiment at LLNL went to General Atomics to head up the DIII-D experimental team; many of the LLNL mirror scientists were assigned to General Atomics to work on tokamaks; Melvin Gottlieb, long-time director of PPPL, retired and was replaced by Harold Furth; Ron Parker succeeded Ron Davidson as Director of the MIT Plasma Fusion Center; Al Trivelpiece left DOE to become executive director of the American Association for the Advancement of Science; N. Anne Davies became deputy director to John Clarke, head of the DOE Office of Fusion Energy (OFE); John Willis replaced Anne Davies as director of the Magnetic Confinement Systems Division in OFE; ICF pioneer John Nuckolls became director of LLNL, to name a few changes that had varying impacts on the directions of the US fusion effort.

ARIES

In February 1988, the US fusion program initiated preconceptual designs of possible tokamak fusion power plants. The designs, under the direction of Bob Conn, and later Farrokh Najmabadi, at UCSD, went under the name of ARIES (Advanced Reactor Innovations Evaluation Study). The study was initially planned to last 3 years, but due to its success and usefulness in guiding the fusion research effort, a series of designs were completed over the next decade.

Congressional Testimony

In April 1988, I presented the following testimony on the ($360 M) FY 1989 fusion budget request to a hearing of the House Appropriations Subcommittee on Energy and Water Development:

> "The effort to develop a fusion energy source has two important and equally essential parts. The first is, in the near-term, to demonstrate fusion plasma conditions of density, temperature and confinement time close to what is required in a fusion reactor. This effort includes developing the necessary technology and understanding for those near-term achievements. The second important and essential element of fusion development consists of those programs required to ensure that fusion will be useful for some practical purpose in the long-term. These programs include concepts and concept improvements whose environmental, safety and economic features will make fusion a winner in comparison with competing technologies. Such efforts include the necessary systems studies and analysis, materials and technology development, new concept experiments, and fundamental physics and engineering research at national laboratories, industries and universities."

> "Maintaining the necessary balance between these essential near-term and long-term aspects of fusion development has become especially difficult for both the magnetic and inertial fusion programs because of the decline in their budgets in recent years and the flat funding projections of the next several years."

"In magnetic fusion, the near-term program consists primarily of the TFTR and CIT tokamak programs at Princeton, other large tokamak programs at the DIII-D at GA Technologies and the Alcator program at MIT, and supporting theory and technology efforts. I would also include in this category the ITER design effort, although the payoff might well be categorized as mid-term. These programs are clearly essential to fusion progress and deserve a certain degree of priority in allocating funds."

"It is unfortunate that the funding projected to be available for the CIT project does not permit it to be built expeditiously. As I warned in my testimony last year, 'unless substantial increases are provided in FY 1989, I would anticipate schedule slippages and project cost increases.' The budget submitted does not provide the increases required for expeditious completion either for CIT or for several other projects, notably the CPRF (reversed field pinch) at Los Alamos and the Field Reversed experiment at Spectra Technology."

"Of more concern than the FY 1989 submission, however, are the out-year projections. The OMB and DOE appear to have agreed on a 5-year projected flat budget profile (for fusion) that, if implemented, not only results in delays and cost increases for CIT but also spells disaster for many, if not most, of the programs required for the success of fusion in the long-term."

"Systems studies, necessary for guiding the magnetic fusion program toward practical products, are projected to decline to a paltry $2 million per year. These studies include the investigation of concepts and fuel cycles that would impact the environmental, safety and economic characteristics of fusion devices. Also, a recent subpanel of the Magnetic Fusion Advisory Committee warned in its interim report that the program currently in place and projected would not result in sufficient data by the year 2005 to permit an assessment of the environmental, safety and economic aspects of fusion to be made nor to allow a fusion demonstration reactor to be built using low-activation materials."

"It is not unreasonable to ask the long-term payoff portions of the fusion program to tighten their belts a little in FY 1989 to help launch a major new near-term project like CIT. It is unreasonable, however, for the Executive and Legislative Branches to expect to build a $400 M new facility (CIT) without providing new money for a substantial fraction of its cost. A new project of this magnitude requires some real budget growth. This problem is not too acute in FY 1989 but will become very acute in later years if the 5-year projections of DOE and OMB are adhered to."

"A similar problem has presented itself in the case of the inertial fusion program. This program has had good experimental success in the last few years and is now ready to begin design leading to construction of a 'Laboratory Microfusion Facility.' Unless real growth in funding is forthcoming, design and near-term R&D for such a facility will restrict our ability to fund those aspects of the inertial fusion program that have long-term civilian payoff. These programs include fusion reactor conceptual design studies and the development of laser and particle beam technologies that are capable of the high efficiency and repetitive pulsing necessary for a commercial power system."

"Of immediate concern to this Committee should be the failure of DOE to request funds for the $27 million ILSE heavy ion fusion accelerator at the Lawrence Berkeley Laboratory. Heavy ion fusion accelerator development has been a part of DOE's Basic Energy Sciences (BES) program. As a consequence of DOE deciding not to fund the ILSE accelerator, they are also seriously considering abandoning heavy ion fusion accelerator development altogether. I urge this Committee to authorize the ILSE project and to direct the DOE BES office to continue this important program."

"Fusion research and development is important to our nation's future. The fusion community appreciates the support it has received from this Committee over the years."

Although government funding for fusion was not adequate for the expeditious development of fusion as an energy source, still the research on plasmas that was fundamental to the fusion process was valued for its own intrinsic value and for the many commercial "spin-offs" for which the scientific underpinnings were being provided by fusion researchers (see Chap. 11).

Major Achievement: Inertial Fusion

In spring 1988, scientists at the University of Rochester achieved compression of deuterium–tritium capsules to 100–200 times liquid density using direct drive and only 2 kJ of laser light [57]. Approximately 700–1,000 times liquid density is required for fusion power plants. Scientists at the Lawrence Livermore National Laboratory had previously achieved similar compressions with indirect drive using 4.5 kJ of laser light from the Novette laser. In direct drive, the laser light is used directly on the surface of the capsule; in indirect drive, the capsule is contained inside tiny cylindrical chamber ("hohlraum") inside which the laser light is first converted to X-rays which drive the compression.

The Arrival of Robert Hunter

In fall 1988, the Senate confirmed the appointment of Dr. Robert Hunter, brother of California Congressman Duncan Hunter, to become DOE Director of Energy Research, to succeed Al Trivelpiece who had resigned 1 year earlier. Hunter was a plasma physicist who had received his Ph.D. from the University of California at Irvine. He had worked at both the Air Force Weapons Laboratory and Maxwell Laboratories before forming his own company, Western Research Corporation. His special technical interest was in krypton–fluoride laser technology.

About the same time, 18 senators, led by Sen. Tim Wirth (D-CO), introduced a new energy policy bill (S.2667) "to establish a national energy policy to reduce global warming, and for other purposes." The bill would have required the Secretary of Energy to provide within 1 year a plan showing how fusion research and development could be brought to fruition by the year 2010.

During the first week of August 1988, the Magnetic Fusion Advisory Committee (MFAC) joined by 30 members of the fusion community held a Summer Study at Coolfont, VA. MFAC chairman Fred Ribe of Los Alamos transmitted the results of the study to Hunter on September 8. In his transmittal letter, Ribe stated, "At present, the U.S. base (fusion) scientific and technology programs are operating at budget levels too low to permit them to support effectively the CIT and ITER programs and to exploit new ideas enhancing the commercial acceptability of fusion." Ribe said the MFAC "strongly supports (the CIT) initiative within a balanced US fusion program." The Summer Study report urged three program thrusts:

> "(1) The US program to construct a compact ignition tokamak (CIT); (2) the international design of an International Thermonuclear Experimental Reactor (ITER); (3) R&D programs to ensure realization of the environmental, safety, and economic potential of fusion as a commercial energy source."

As 1988 drew to a close, CIT construction planning was proceeding apace. Princeton awarded a multiyear contract for much of the CIT to an industrial team headed by Ebasco Services, Inc. The team included McDonnell Douglas Astronautics Company and Spar Aerospace, Ltd. Other industrial subcontractors included SAIC, Telerobotics International, Remotec, VetcoGray, and Process Applications, Inc.

DOE's Robert Hunter had other ideas, however. During fall 1988, Hunter decided to shift emphasis within the magnetic fusion program to put more effort on near-term studies of tokamak physics and less emphasis on long-range technology, materials research, alternate and advanced concepts, and preparation for tritium operation in TFTR [58]. An immediate casualty of Hunter's action was that approximately 120 industry personnel were given 1-day's notice to stop work at the Princeton Plasma Physics Laboratory. Hunter's action also resulted in an additional 2-year slippage in the planned date for deuterium–tritium operation in TFTR. Hunter also charged MFAC to carry out a study on "confinement and machine physics" to summarize "the status of our understanding of confinement in tokamaks and recommend what needs to be done to improve understanding." Hunter set up an MFAC subpanel to carry out this study under the chairmanship of Kim Molvig (MIT), with a reporting date of April 1989.

In January 1989, DOE's John Clarke was given an "indefinite leave of absence" from his post as head of the Office of Fusion Energy in order to head a working group on climate change and to chair a task force "to address the issue of how DOE's energy research programs should integrate into the Intergovernmental Panel on Climate Change." Clarke's deputy, N. Anne Davies, was named Acting Associate Director for Fusion Energy. Although Clarke was expected to return to his fusion post at a later time, in fact he never did return. In January 1989 also, George H. W. Bush succeeded Ronald Reagan as US President. He named retired Admiral James Watkins to be Secretary of Energy. Watkins had a strong nuclear engineering background and had commanded a nuclear submarine. The FY 1990 budget submitted to Congress in January 1989 showed fusion holding level at $390 M, with the CIT proposed to also hold its FY 1989 funding of approximately $20 M.

The CIT Fiasco

In early March 1989, the MFAC subpanel, chaired by Kim Molvig, issued a lukewarm assessment of "the certainty of achieving ignition in the proposed CIT." The subpanel concluded "the overall expectation for ignition in the first phase (of CIT operations) is very low." They noted that by increasing the magnetic field and plasma current and also adding additional heating power, "the CIT would have a high probability of ignition in the second phase." "However," the subpanel said, "without a physics basis for the empirical scaling laws, and without known scaling behavior of the enhanced regimes, there cannot be confidence at the present time in obtaining the performance required for ignition." The full MFAC cautioned DOE not to equate CIT with only its first phase and endorsed its continued construction.

At a Senate hearing on June 14, 1989, Hunter proposed that Congress cut the President's FY 1990 fusion budget request by $50 M. Alerted to Hunter's proposal, the House Appropriations Committee voted to cut the fusion budget request by $68 M, including the $18 M requested for continuation of CIT construction. On June 15, Energy Secretary Watkins sent a letter to the various committees of Congress stating, "Due to lack of resolution of key scientific unknowns, the Department is no longer proposing to begin actual construction of this device (CIT) in FY 1990." In his letter, Watkins also stated, "After receiving the formal report of the (MFAC) review panel, I met with leaders of the magnetic fusion community in late April to discuss (the panel's) findings. I am now convinced of the possibility that the CIT will be unlikely to meet the key ignition goal. . . ." In his letter, Watkins also indicated his intention to conduct a "high level policy review" of the fusion program logic. On June 21, the MFAC took the unprecedented step of writing directly to Secretary Watkins. In the letter, signed by its membership unanimously, the MFAC said, "We take exception to the CIT being equaled to only its first phase; doing so conceals the sound logic and value of the CIT project." They pointed out that "in order to achieve ignition at minimum cost, the project was laid out with two phases, only the second of which was regarded as having high probabilities of ignition." The Congress would eventually cut the President's FY 1990 budget request of $390 M to $317 M, and the CIT was gone, never to return.

In a September 12, 1989 speech on the floor of the US House of Representatives, Rep. Robert Roe, chairman of the House Committee on Science, Space, and Technology, said, "I do not think it is the intent of Congress to give to a Secretary, no matter who he is, the total right to determine the fusion program of this country after the taxpayers have spent $6 billion on that program to date." He scheduled three half-days of hearings on fusion, Oct 3, 4, and 5. According to Roe, "The purpose of the hearings is to assess the progress of the Magnetic Fusion Energy Program, and to evaluate the Department of Energy's plans for carrying this program forward." Twelve witnesses, listed in Table 6.1, testified the first 2 days, followed by DOE's Robert Hunter on the third day. Of the 12 witnesses, 11 favored proceeding with CIT and one (Molvig) favored delay. In a surprise move (welcomed by the fusion community), Hunter left DOE at the close of business on October 26 and returned to the private sector. Jim Decker was once more named Acting Director of Energy Research.

Table 6.1 Witnesses appearing at the October 3–5, 1989 congressional hearings

Harold Furth (Princeton Laboratory)	Kenneth Gentle (University of Texas)
John Gilleland (Livermore Laboratory)	Robert Iotti (Ebasco Services)
Alexander Glass (Livermore Laboratory)	Kim Molvig (MIT)
Rulon Linford (Los Alamos Laboratory)	Fred Ribe (University of Washington)
Ronald Parker (MIT)	John Sheffield (Oak Ridge Laboratory)
David Overskei (General Atomics)	Robert Hunter (US Department of Energy)
Stephen O. Dean (Fusion Power Associates)	

The experience of the US fusion program during Robert Hunter's 1-year tenure at DOE was summarized in Fusion Power Associates November 1989 Executive Newsletter as follows:

> "During his 1-year tenure, Hunter has been the nemesis of several energy research programs, including the fusion and high energy physics programs. The problems of the magnetic fusion program under Hunter's direction have been legion and have descended upon the program regularly without advance warning."

> "Early during Hunter's tenure, word began to reach the fusion community that he was denigrating the progress of the program during visits with Congressional staff. In October 1988, in a banquet speech to the American Nuclear Society fusion engineering conference, he stated that no progress had been made in 10 years and that he found no support for fusion on The Hill. About that same time, he decided to "reprogram" $23 million of magnetic fusion funds. This action resulted in the layoff of 160 industry personnel at the Princeton Plasma Physics Laboratory on 1-day's notice and caused major disruption of several programs, including ATF at Oak Ridge and CPRF at Los Alamos. He quoted the results of the MFAC 1988 Summer Study as a partial justification of the reprogramming, but MFAC members felt he had inappropriately focused on only a portion of the Summer Study report."

> "Hunter then removed the director of the U.S. fusion program, John Clarke, and during all of 1989 that program has been without a permanent director. Hunter set up a shadow management, consisting of two outside consultants (Tom Johnson from West Point and Kim Molvig from MIT), to plan, manage and oversee magnetic fusion. Johnson created a radical new fusion policy document without consulting any of the DOE staff charged with managing the program under the able leadership of Dr. N. Anne Davies. All budgets and program actions proposed by Dr. Davies and her staff were reviewed and approved by Dr. Molvig and Hunter personally. The new policy was being used to shape the fusion program since early spring but its existence was not publicly known until May when a copy leaked. Even then, Hunter refused to acknowledge its authenticity."

> "Early in the year, Hunter asked MFAC to review the status of tokamak confinement physics, including the "probability" of ignition in CIT. He asked MFAC to appoint Dr. Molvig to chair the panel. Based on one section of the report, eventually discounted by the full MFAC, Secretary Watkins decided not to proceed with the CIT project and secretly endorsed Hunter's new policy, which he called "new" and "innovative." Watkins prepared to announce this policy (including a $50 million cut in magnetic fusion, postponement of CIT, and establishment of a civilian energy competition with inertial fusion) at a Senate hearing in June. At the last minute, under pressure from several Senators, Watkins pulled back from endorsing the plan and instead agreed to a "high level independent review" of the "draft policy." He did, however, send letters to the Hill requesting that CIT construction be postponed."

Major Successes: DIII-D and JET

In the summer and fall of 1989, scientists at General Atomics using the DIII-D tokamak achieved record values of plasma pressure relative to magnetic field pressure of 9.3%. Values in the range of 8–15% are believed necessary for a commercial tokamak fusion power plant.

In a press release November 7, 1989, and presented November 13 at the American Physical Society Division of Plasma Physics meeting, scientists working on the Joint European Torus (JET) in the UK reported achieving record values for the fusion equivalent breakeven parameter (Q). Q is the ratio of fusion energy produced to energy in the plasma. They achieved equivalent Q values in the range of 0.7–0.8 in deuterium plasmas. "Equivalent Q values" refer to the fact that the results were achieved in deuterium plasmas rather than the eventually required mixture of deuterium and tritium. The JET team leader stated in the press release, "I consider that the problems of heating and confining a thermonuclear plasma are now solved." He said, "Over the past year the JET project has more than doubled its best fusion factor (Q) and has now basically achieved its principal objective of establishing the scientific feasibility of nuclear fusion as an energy source."

More on Inertial Confinement Fusion

The inertial confinement fusion program, funded within the DOE's weapons budget category, made significant progress during the 1980s and was being taken ever more seriously as a potential competitor to magnetic confinement fusion as a possible future energy source. In mid-1986, the White House Office of Science and Technology Policy (OSTP) issued the report of an ICF review panel that stated, "Based on its work to date, the Committee is convinced that the ICF program is a vigorous and successful research effort which has made striking progress over the past few years." The committee stated, "ICF may eventually lead to commercial power." Shortly thereafter, the House Science and Technology Committee recommended an additional $3 M be added to the President's budget request, saying, "In Fiscal Year 1987, the Committee directs the Department of Energy to investigate the civilian application potential of inertial confinement fusion in view of the terrestrial (above ground) progress achieved in the ICF laser program as well as in complementary underground tests (House Report 99–719, Part 1)." In mid-1987, ICF scientists began planning for a major facility called the Laboratory Microfusion Facility (LMF) that would, if successful, produce fusion yields well beyond breakeven conditions. I headed up a study for DOE [59] on the "Status of Candidate Drivers for a Laboratory Microfusion Facility" and presented an invited talk on the subject at the 8th Topical Meeting on the Technology of Fusion Energy in Salt Lake City on October 10, 1988. In May 1989, the DOE published a 95-page report (DOE/DP-0069) titled "Laboratory Microfusion Capability (LMC) Study Phase I Summary."

As indicated earlier, in the spring of 1986 scientists at the Lawrence Livermore National Laboratory, using the new Nova laser, achieved the Lawson density–confinement time parameter (nτ), reaching a value of 2×10^{14}. The temperature achieved was 1.5 keV, only a factor of about 3 below the Lawson ideal ignition temperature of 5 keV. One year later, the Nova laser was used to achieve a 30-fold capsule compression ratio. Compression ratio is the ratio of initial to final capsule radius. A value of about 40 is believed to be necessary for high gain for fusion power plants.

The traditional method is to place the capsule inside a small cylindrical chamber called a hohlraum and aim the laser beam into the hohlraum through two tiny apertures at either end, convert the laser energy into X-rays inside the chamber, and use the X-rays to implode the capsule. For fusion power plants, laser implosions would be more efficient if the laser energy could be used directly on the surface of the capsule without a hohlraum. This technique was called "direct drive" to differentiate it from the hohlraum technique called "indirect drive." In order to do direct drive without driving instabilities during the implosion process required the development of techniques to "smooth" the laser beam profile. Two techniques were successfully developed (at the Naval Research Laboratory and at the University of Rochester) to do this. Rochester scientists also demonstrated in 1987 radically increasing the intensity of a laser pulse to the terawatt level, using a technique call "chirped-pulse amplification." These pulses were 1,000-fold more intense than previously possible. In spring 1988, scientists at the University of Rochester, using direct drive, succeeded in compressing deuterium–tritium capsules to 100–200 times liquid density using less than 2 kJ of laser light. Similar compressions using indirect drive at Livermore had required about 4.5 kJ. What was more important, in the view of the Rochester team, was the high degree of spherical symmetry that was achieved in the experiments as a result of the "smoothing" of the laser beam.

In spring 1989, in response to a congressional mandate, the DOE set up an ICF review under the auspices of the National Academy of Sciences. The academy was asked to review the accomplishments of the ICF program and, for the first time, asked to evaluate the "civilian energy potential of this technology." Dr. Steven Koonin of the California Institute of Technology was named to chair the review. Meanwhile, Robert Hunter, as part of his plan to restructure the magnetic fusion program, as discussed earlier, announced his desire to take $50 M from the magnetic program and use it to initiate a civilian inertial fusion energy program.

In September 1989, the DOE Inertial Fusion Division (within DOE's weapons program) issued an unclassified version of a 5-year plan [60]. In the preface, division director Sheldon Kahalas states, "the purpose of the FY 1990–1994 Inertial Confinement Fusion (ICF) Program Plan is twofold":

"First, it indicates that there are new (1987 and 1988) ICF experimental results that are of historical significance. These new results provide increased confidence in the technical feasibility of ICF and have motivated the Inertial Fusion Division (IFD) to begin planning for a Laboratory Microfusion Facility (LMF). Such a facility could begin to provide a significant return on the investment in ICF research and development in the form of both military and civilian applications at about the turn of the century."

"Second, it presents a strategic plan for accomplishing this LMF objective. The plan presents a range of resource options consistent with achieving the objective of an LMF on a nearer or longer term schedule, depending on the policy selected."

Fusion Policy at the Department of Energy

As the decade of the 1980s came to a close, the fate of fusion at the DOE was anyone's guess. Energy Secretary Watkins had not-so-quietly removed Robert Hunter and promised that he would authorize an independent fusion review. However, the CIT had been canceled, and as a result of the confusion, Congress eventually appropriated $284 M for magnetic fusion in FY 1991, substantially less than the program had in either FY 1989 ($345 M) or in FY 1990 ($317 M).

DOE was preparing a national energy strategy under the direction of DOE Deputy Undersecretary for Policy, Planning and Analysis Linda Stuntz. However, speaking to a meeting of the DOE Energy Research Advisory Board (ERAB), Stuntz said [61], "There's not exactly a huge public sector out there when it comes to fusion programs and other science programs at DOE."

In an interview with reporter Keay Davidson, published in the October 29, 1989 issue of the San Francisco Chronicle, Secretary Watkins is quoted as saying the "(fusion) track record hasn't been all that impressive. I was at Oak Ridge National Laboratory in 1958 and I was told by all the scientists we were within 7 years of fusion . . . for sure by 1965! Well, here we are in 1989 and people are telling me, 'we're just right there, all we need is (to generate) a little more magnetic field and we'll have a wonderful time.' That's one way to do it—but I don't believe it's the best way now, not with the (federal) budget crunch." According to the chronicle, Watkins said that solar is "going to get a lot more support from me," adding, "we're going to be very serious about conservation and renewables, with photovoltaics right up there in front." And speaking to the National Coal Council on November 9, Watkins said, "I can assure you that under my watch the influence of your expertise will continue," adding, "rest assured that this Administration will see the clean coal technology program through to the full extent of its $5 billion, 5-year commitment."

Hope for Resurgence: 1990–1995

7

Abandon Certainty! That's life's deepest command.
That's what life is all about. We're a probe into the unknown,
into the uncertain.

–Frank Herbert
Author, Children of Dune

Input to Energy Secretary James D. Watkins

Energy Secretary Watkins began preparation of a national energy strategy with a meeting on January 11, 1990, in Honolulu. Two witnesses presented testimony on fusion: Harold Forsen (Bechtel Group, Inc.) and Akira Hasegawa (Bell Telephone Labs) [62].

Forsen testified, in part: "We should move forward with a program that will tell us the science of burning plasmas such that the technology development necessary to support that science in fusion power plants can be identified and undertaken on a program schedule to match the need". He said, "We cannot say for sure today what the economics of fusion power might be. It does seem clear, however, that its emissions, safety and radiological inventory have the potential to be more attractive than the nuclear plants we have today". With respect to international collaboration, which he termed "reasonable" and "good," he said "It is important to add, however, that any complacency by the United States to not have a strong domestic program in light of where the international program is headed, is sure to lead to commercial disaster for us as suppliers of goods and services in the potential deployment of these systems. An important point here to add is that the United States program lacks sufficient industrial participation as compared with, say, the Japanese". He added, "Mechanisms and budgets must be found to bring U.S. industries into our program in a meaningful way such that engineering systems and hardware requirements are understood by those who will ultimately build these plants and guarantee their performance".

Hasegawa testified, in part: "Our strong dedication to this goal (of developing fusion as an energy source) has been further intensified by the recognition that the

S.O. Dean, *Search for the Ultimate Energy Source: A History of the U.S. Fusion Energy Program*, Green Energy and Technology, DOI 10.1007/978-1-4614-6037-4_7,

nation and the world are facing a new kind of an energy crisis, one brought about by the deleterious effects of fossil fuel energy sources on the environment. For the longer term, it appears to us that the development of the more benign energy sources represented by improved fission reactors, solar, and fusion is essential and urgent and needs to be a strong element in the National Energy Strategy being developed by your Department".

In the February 1990 issue of the Fusion Power Associates Executive Newsletter [62], and sent to Secretary Watkins, I editorialized as follows:

"The world will soon need new central station power plants for electricity and increasing amounts of fuel for the transportation sector. However, once-vigorous energy programs have been cut to subcritical funding levels because low oil prices and surplus energy supplies during the 1980s have created a false sense of energy security. In addition, demands on the federal budget, combined with a large federal deficit, present a seemingly insurmountable obstacle to obtaining the funds necessary for the timely development for long-range energy technologies".

"A possible solution to the above dilemma would be to create an Energy Technology Development Trust Fund, analogous to the Highway Trust Fund. The Highway Trust Fund collects money from those who use highways now, to pay for providing adequate highways in the future. So too, the Energy Trust Fund would be funded by a set-aside from fossil fuel revenues to be used to ensure that we have technologies to replace fossil fuels in the future".

"In the ideal world of the twenty-first century, transportation will be mainly provided by electric vehicles, while electricity will be generated by non-fossil fuel technologies. The key non-fossil fuel technologies are fusion, fission and various "renewables," including solar. To ensure the availability of these technologies for central station power early in the twenty-first century requires development programs funded at approximately \$5 billion per year. However, the U.S. Department of Energy currently funds these programs at about \$1 billion per year. In the U.S. there are approximately \$200 billion per year in sales of fossil fuel generated electricity and an additional \$200 billion per year in gasoline sales. Consequently, a national sales tax on electricity and gasoline of about 1% would provide the revenues necessary for the Energy Technology Development Trust Fund".

"There are a number of important issues related to establishing such a fund. These include the legal framework and management structure of the organization charged with allocating the funds, and ensuring that the funds are used for the development of long-range technologies and not diverted to near-term fixes. However, a thorough analysis and public discussion of these issues should result in an acceptable charter for the establishment of such a fund. The principle on which such a fund should be established is that the users of a scarce national resource (e.g., fossil fuels) should be required to set aside a portion of current revenue to develop replacement technologies for a time when this resource is no longer available or not socially acceptable".

Around the same time, Richard P. Hora and S. Locke Bogart (General Dynamics Space Systems Division) published [63] an article proposing that a similar "amortizing fund" be established to accumulate funds to support R&D aimed at advanced electric generating systems.

Table 7.1 DOE 1990 Fusion Policy Advisory Committee (FPAC)

H. Guyford Stever, (National Academy of Engineering), Chair	Willliam Herrmannsfeldt (Stanford Linear Accelerator Center)
Roger Batzel (Lawrence Livermore National Laboratory, retired)	Arthur Kerman (MIT)
Ira B. Bernstein (Yale University)	Kenneth L. Kliewer (Purdue University)
Robert W. Conn (UCLA)	John W. Landis (Stone and Webster Engineering Corporation)
E. Linn Draper (Gulf States Utilities)	R. Bruce Miller (Titan Corporation)
Harold K. Forsen (Bechtel Group, Inc.)	Barrett H. Ripin (US Naval Research Laboratory)
John S. Foster, Jr (TRW, retired)	Marshall N. Rosenbluth (University of California, San Diego)
T. Kenneth Fowler (University of California, Berkeley)	Robert Sproull (University of Rochester, Emeritus)
Melvin B. Gottlieb (Princeton Plasma Physics Laboratory, retired)	Richard Wilson (Harvard University)

Fusion Policy Advisory Committee Review Begins

On March 19, 1990, Energy Secretary Watkins announced the initiation of his promised comprehensive fusion policy review. He appointed Dr. H. Guyford Stever to chair the 18-member Fusion Policy Advisory Committee (FPAC). Stever had been head of the White House Office of Science and Technology Policy office under President Gerald Ford, a former Director of the National Science Foundation and President of Carnegie Mellon University.

In announcing the review, Watkins stated, "Fusion energy has the potential to provide electricity in the next century safely and without the creation of highly radioactive waste products or greenhouse gases. The development of fusion as a viable technology is an important long-term goal of the Department". The announcement stated that the FPAC report (to be completed by September 1990) should provide advice on how to structure the department's magnetic and inertial confinement fusion programs, including advice on the balance of research activities within the programs, the timing of experiments to test the burning of plasma fuel, the International Thermonuclear Experimental Reactor (ITER), and the development of laser technologies. The makeup of the FPAC is shown in Table 7.1.

At its first meeting, March 22–23, I provided the FPAC the following fusion policy statement from the Fusion Power Associates Board of Directors:

The Board of Directors of Fusion Power Associates urges that fusion development be based upon the following recommended policies:

- Timely commitments to necessary new and improved experimental facilities are needed to ensure continued program momentum and progress into the 1990s.
- It would be premature, at this stage, to judge which of the variety of magnetic and inertial fusion concepts will ultimately succeed commercially. This fact

should not discourage use of the best available concepts in the design and construction of needed fusion test facilities.
- International cooperative agreements have been a substantial factor in fusion progress and should be encouraged. Such agreements, however, are not an effective substitute for focused national efforts, needed national experimental facilities, and subsequent engineering test devices.
- Governments should encourage and promote the full participation of industry in the planning, research and development, as well as in engineering and operating aspects of fusion programs. If industry participates in the government-funded R&D programs now, industry will be better prepared to assess the commercial potential of fusion power in the future.
- The time has come to place increased emphasis during the next decade on the engineering and systems design aspects of practical fusion energy systems.

Bush–Gorbachev Summit Meeting

US President George H. W. Bush and Soviet Premier Mikhail Gorbachev held a summit meeting in June 1990, after which they issued the following joint statement on fusion:

> "At their meeting in Geneva in 1985, the leaders of the United States and the Soviet Union emphasized the importance of the work aimed at utilizing controlled thermonuclear fusion for peaceful purposes, and advocated the widest practical development of international cooperation in obtaining this essentially inexhaustible source of energy for the benefit of all mankind".
>
> "The International Thermonuclear Experimental Reactor (ITER) project, involving joint efforts by the USSR, the United States, Japan and the European Community, under the aegis of the International Atomic Energy Agency, is making significant progress towards this end. A conceptual design will soon be completed".
>
> "Noting with satisfaction the results attained under this project, the United States and the Soviet Union look forward to continued international efforts aimed at promoting further progress in developing controlled thermonuclear fusion for peaceful purposes".

PBS Film: Fire from the Sun

Starting June 3 and throughout summer 1990, public television stations around the country aired a new fusion documentary *Fire from the Sun: The Search for Fusion Energy,* a one hour film produced by Michael Pack (Manifold Productions), hosted by E. G. Marshall. The film "places the 40-year effort to master fusion in a larger political and cultural context". Many fusion scientists, including myself, presented their views in the film. The film is available on DVD from Manifold Productions [64].

FPAC Reports

In September, the FPAC provided its report [65] to Energy Secretary Watkins. The FPAC made the following Policy recommendations:

- The United States should commit to fusion as a potential energy source.
- The United States fusion program must become energy oriented, with specific goals of an operating demonstration power plant in 2025 and an operating commercial power plant by 2040.
- The United States should take maximum advantage of international collaboration while maintaining a sound domestic program.
- Involvement of the private sector should be an integral part of the energy-oriented fusion program.

With respect to strategy, the FPAC stated:

"Both MFE and IFE should be supported, although the two are at different levels of maturity, and have different technical pathways toward achieving their goals. In recommending this strategy the Committee affirms its belief that the two approaches to civilian fusion energy are not ready for a choice of one over the other. While MFE is closer to achieving its scientific goals, the physical separation of driver and fusion reactor may ultimately be an important advantage for IFE. Pursuing both options at this time reduces technological risk. It is possible that a choice will take place prior to the construction of the Demonstration Power Plant, and perhaps earlier".

"The strategy for both MFE and IFE is first to attain sufficient understanding, including experimental verification of the scientific principles and phenomena related to fuel-burning plasmas and ignited pellets, then to demonstrate this understanding in an Engineering Test Facility that includes engineering and technological features, and finally, to move towards application through a Demonstration Power Plant. These demonstrations alone, however, are insufficient. There must be an independent program of concept improvement, including study, and where promising, development of alternative configurations that may be more suitable for commercialization. In addition, the overall program must include vigorous technology and materials development. Specialized fusion technology and materials development are as important to the eventual economic success of fusion as are the demonstrations of physics performance, and some of the development will require specialized facilities".

FPAC also criticized DOE for its "over-classification" of inertial fusion, saying, "Classification policies and restrictions on transfer of fusion technology to foreign nations should be re-examined. Classification hinders the inertial fusion program". They stated, "The Committee believes that classification guidelines can be written to prevent transfer of weapons technology and yet permit collaboration on the processes and targets of interest for energy applications. These changes should be made as soon as possible, before a damaging effect on public acceptance of fusion energy is caused".

Secretary Watkins immediately ordered [66] "a comprehensive review of the classification of our programs in inertial confinement fusion". However, 1 year later the DOE had still not completed its classification review and it would fall to Watkins successor, Hazel O'Leary, to complete this process 2 years later (in December1993).

The DOE did begin inertial fusion power plant studies in mid-1990. Contracts were given to two industrial teams; one led by McDonnell Douglas Missile Systems Company and one led by W. J. Schafer Associates. KMS Fusion, TRW, UCLA, Ebasco, Spar Aerospace, and Ontario Hydro were on the McDonnell Douglas team; University of Wisconsin, Bechtel, General Atomics, and AVCO were on the Schafer team. These studies were completed in 1992. Although the data provided for the studies by DOE labs had been reviewed by DOE classification officers throughout the process, the final reports got held up by classification officers just as they were about to be issued. I reported on this in the July 1992 Fusion Power Associates Executive Newsletter and, shortly thereafter, was "called on the carpet" by the DOE security office. My meeting turned out not to be with a DOE security/classification official but with one or their industrial support subcontractors. I pointed out that I had not divulged any classified data, but he told me that simply stating that classified data was the cause of holding up release of the reports was a security infraction. I apologized and heard nothing further about the matter. I expect there is still a note on this in my government file somewhere.

When DOE began declassifying much of inertial fusion in mid-1992, they kept the Halite-Centurion data on inertial confinement fusion obtained from underground nuclear weapons tests classified. When the existence of this program was first exposed by William Broad on the front page of the March 21, 1988 New York Times, DOE had responded [67] only that "Underground tests in Nevada have allowed demonstration of excellent performance, putting to rest basic questions about the feasibility to achieve high gain".

In the fall 1992 issue of the National Academy of Sciences magazine, *Issues in Science and Technology,* Edward Teller, a pioneer of both the US weapons and civilian fusion programs, said, "Our keeping secrets has often misled and confused our own people but has been ineffective in denying information to our enemies or competitors. I make a proposal, hoping that it may help to start a fruitful discussion. Let us pass a law requiring all secret documents to be published 1 year after their issuance".

Budget Cut

The FPAC recommended FY 1992 funding for MFE of \$420 million, compared to the President's FY 1991 request level of \$325 million (which had not yet been passed by Congress). One month after FPAC issued its report, Congress passed the FY 1991 MFE budget, cutting it to \$275 million. The Congress did not cut the President's \$175 million request for inertial confinement fusion.

Secretary Watkins would successfully convince Congress to allow him to "reprogram" \$25 million back into the MFE program from other parts of the DOE budget a few months later. However, immediately after the initial \$50 million cut was signed into law, DOE fusion program management took action to eliminate \$50 million of non-tokamak fusion research activities. These actions included termination of the Advanced Toroidal Facility (ATF), a new stellarator at the Oak Ridge National

Laboratory; a new Reversed Field Pinch facility, nearing completion at the Los Alamos National Laboratory; a Field Reversed Concept program at Spectra Technology; and a Spheromak program at the University of Maryland.

The MFE program cuts instituted by DOE marked the beginning of a several year policy to eliminate "alternate" (non-tokamak) concepts in order to focus program resources on the tokamak concept. Secretary Watkins sought to reauthorize construction of the Compact Ignition Tokamak (CIT) during preparation of the FY 1992 President's budget but his request was denied by the Office of Management and Budget. The President's FY 1992 budget request did, however, contain $30 million for "R&D, prototype fabrication, and design for the Burning Plasma Experiment (BPX)." BPX was the name given to the formerly named (but more ambitious) CIT. The FY 1992 budget request also established a new category for Inertial Fusion Energy (IFE) within the largely-tokamak DOE Office of Fusion Energy (OFE). The new IFE category consisted largely of a Heavy Ion Fusion effort that had previously been funded as accelerator development within DOE's Basic Energy Sciences category.

The Boston Globe, in its January 2, 1991 edition, editorialized: "Among the long-term solutions to the world's energy problems, none is quite as enticing as fusion power—an elusive objective that will demand enormous scientific and technological effort to succeed. At a time when others are accelerating their commitments, Congress has unwisely decided to curtail funds for the American program—a move that amounts to shortsighted miserliness".

An Accelerated Fusion Power Development Plan

At Fusion Power Associates annual meeting and symposium "Energy for a New Age," held in September 1990, I presented the results of a 1-year study [68] by a small group I had assembled in 1989, consisting of Charles Baker Daniel Cohn, Susan Kinkead, and myself. The paper, entitled "An Accelerated Fusion Power Development Plan," aimed at operating a fusion "pilot plant" within 15 years of a commitment to proceed. The report stated, "Because the plan is ambitious and based on a tight time schedule, it entails considerable technical risk. The initiative controls this risk by requiring an intense up-front period (7–8 years) of project-specific R&D and plant design". Our group, with several additional authors, went on to produce several more papers on pilot plant concepts [69–71].

1991 National Energy Strategy

On February 20, 1991, Secretary Watkins released the Department of Energy's long-awaited National Energy Strategy. The report states, "Fusion is an important, albeit long-range, element of the National Energy Strategy because of its many potential advantages as an energy source". It notes, "The technical complexity associated with fusion development is such that substantial investments are required for new experiments, design (of) facilities, and test facilities. This implies the need for long-term growth in research and development funding". The report sets as a "goal" to "prove fusion energy to be a technically and economically credible energy source, with an operating demonstration plant by about 2025 and an operating commercial plant by about 2040". This goal is set for both the MFE and IFE programs. The report states, "In the interest of achieving this goal, actions will be undertaken to encourage substantial involvement of U.S. industry in fusion energy development. This participation will be sought not only for the hardware phases of the program, but also in the planning, research and development, and analytical phases".

New Fusion Energy Advisory Committee

In June 1991, DOE named N. Anne Davies associate director for fusion energy (she had been in an "acting" capacity since 1989) and established a new Fusion Energy Advisory Committee (FEAC) to replace the MFAC, which had been functioning during the 1980s. Bob Conn (UCLA) was named chairman. The members of FEAC are shown in Table 7.2. The FEAC would carry out three reviews via a set of panels in the 1991–1992 time period. The first (chaired by Rulon Linford) considered the appropriate scope and mission of ITER. The second (chaired by David Baldwin) considered post-TFTR initiatives in the absence of BPX. The third (chaired by myself) considered the current policy on fusion alternate concepts and concept improvement.

Based on the first panel report, FEAC chair Bob Conn told (new) DOE Director of the Office of Energy Research William Happer, "FEAC views ITER and its Engineering Design Activities (EDA) phase as a central element of the U.S. magnetic fusion program. Further, we strongly reaffirm the importance of integrated nuclear testing as a key part of the ITER mission". Conn states, "The necessity of using ITER for the first detailed investigations of high-Q and ignited burning plasmas will extend the phase of ITER dedicated mainly to such physics issues. This first phase is now estimated to take as much as 10 years". The letter says, "Additional complementary activities dedicated to acquiring part of the nuclear testing data would permit shortening the ITER test program. FEAC recommends that a study of the feasibility of such a complementary program be undertaken".

Table 7.2 Members of DOE Fusion Energy Advisory Committee (FEAC), 1991

Robert Conn (UCLA), Chair	Bob McCrory (U. Rochester)
Dave Baldwin (LLNL)	Norman Ness (U. Delaware)
Klaus Berkner (LBNL)	David Overskei (General Atomics)
Floyd Culler (EPRI)	Ron Parker (MIT)
Ron Davidson (PPPL)	Barry Ripin (NRL)
Stephen O. Dean (Fusion Power Associates)	Marshall Rosenbluth (UC, San Diego)
Dan Dreyfuss (Gas Research Institute)	John Sheffield (ORNL)
John Holdren (UC, Berkeley)	Peter Staudhammer (TRW)
Rulon Linford (LANL)	Harold Weitzner (NYU)

Based on the second panel report, FEAC chair Bob Conn advised Happer, "to proceed with a Steady State Advanced Tokamak (SSAT) as a FY 1994 construction project". The device is estimated to cost "about $500 million in as-spent dollars," he said. The SSAT would later be renamed the Tokamak Physics Experiment (TPX).

The third FEAC panel report, which I chaired, stated, "A breakdown in communications occurred between DOE fusion program managers and the fusion research community in the fall of 1990 when DOE decided to narrow the program to tokamaks. A negative consequence of that decision is the widespread impression that DOE has postured itself to be unreceptive to new ideas. It is important to reverse this decision". In requesting the Panel 3 review, Happer had stated, "The overall policy question is whether, given the demands of the mainline tokamak program and current budget constraints, we should encourage and fund proposals on concepts other than tokamaks". FEAC chair Bob Conn told Happer, "the tokamak has emerged as the most scientifically successful. DOE's policy should be based on the recognition that tokamak concept improvement programs are essential and should receive highest priority". However, Conn stated, "It is also true that uncertainties remain in the extrapolation of the tokamak to a competitive commercial reactor. As long as such uncertainties remain, a non-tokamak fusion concepts program, as some level, should be supported as a matter of policy".

These FEAC reports were published in the Journal of Fusion Energy [72].

Major Advance: DIII-D

In the summer of 1991, scientists at General Atomics, using the DIII-D tokamak, discovered operating conditions in which the energy of the plasma is contained for almost twice as long as the best results obtained previously in any other tokamak. For many years many tokamaks have been operating in what is called H (for high) mode, which had confinement times about twice the more common L (for low) mode. The new regime, called VH (for very high) mode, is typically 1.8 times H mode.

The scientists were also able, in separate experiments, to achieve ratios of plasma pressure to magnetic field pressure approaching 50%, another record.

Happer at the Helm

In July 1991, President Bush had nominated Professor William Happer of Princeton University as director of the DOE Office of Energy Research. Jim Decker had served as acting director since the departure of Robert Hunter in late 1989. Watkins also announced that Happer would serve as his science and technology advisor. Watkins said that Happer would "help facilitate the transfer of technology from our laboratories to industry". In a July 22 letter [73] to me as president of Fusion Power Associates (FPA), Happer thanked FPA for its support of his nomination and said, "I look forward to discussing with you strategies for mastering magnetic fusion in the most incisive and expeditious way, subject to the budget constraints we are likely to face".

Happer arrived at DOE in August 1991 as the FY 1993 budget proposals were being developed for submission to the Office of Management and Budget (OMB). He found that his budget "guidance" from OMB was that there would be no budget growth for the programs under his jurisdiction with the exception of construction funds for a large new high energy physics accelerator called the superconducting supercollider (SSC). For fusion specifically, there would be no funds available for the construction of the Burning Plasma Experiment (BPX) which had succeeded the CIT in DOE fusion planning. The BPX was estimated to cost $1.4 billion. (The SSC was later cancelled when its construction cost escalated to approximately $10 billion).

To confront this budget dilemma, Happer quickly formed a "Task Force of the Secretary of Energy Advisory Board (SEAB) on Energy Research Priorities". Charles Townes, a Nobel laureate from the University of California, Berkeley, who had coinvented the laser, headed the task force. Two fusion scientists, Marshall Rosenbluth (UC, San Diego) and David Baldwin (LLNL), were part of the 15-member group.

In addition to fusion, the task force reviewed a large number of other projects in the DOE Office of Energy Research, including high energy and nuclear physics. The task force recommended that the separate program advisory committees evaluate the priorities within their respective areas under various budget scenarios. With respect to fusion, the task force stated, "The Task Force believes that funding for the magnetic fusion program must increase at a modest rate (e.g., 5% real growth per year) even at the expense of other programs. This recommendation follows from the opportunity to participate in the International Thermonuclear Experimental Reactor (ITER), as well as a recognition that no major fusion facilities have been authorized since 1976 and many programs have been canceled, so that the domestic program is in danger of no longer being able to fulfill its scientific and educational mission". "Such a modestly growing program is incompatible with authorization of the Burning Plasma Experiment (BPX) especially since the estimated cost of BPX has risen," they said. They added, "Concept exploration should begin to define a new experiment in the $500 million class for the purpose of scientific study of tokamak improvements that could suggest new operating modes for ITER and permit the design of more reactor-desirable follow-ons to ITER".

A meeting of fusion community leaders was held October 16–17, 1991, in San Diego, and a technical workshop was then held October 25 at MIT. The focus of the discussions was to define a major new tokamak facility in the $300–$400 million range that would be designed to take maximum advantage of existing facilities at the Princeton Plasma Physics Laboratory (where BPX would have been located).A range of options were discussed, including copper vs. superconducting coil devices.

A large number of workshops and panels (including a New Initiatives Task Force led by John Sheffield) would be active during the winter and into the spring (1992) to decide what should be the elements of the US magnetic fusion program in the absence of BPX. One of the ideas considered during this period was the need for a "small" (50 MW compared to ITER's planned 1,000 MW) dedicated fusion technology facility to test components. Mohamed Abdou (UCLA) spearheaded this. Twenty years later, he would still be championing the need for such a facility. The FEAC panel considering these issues dubbed the inclusion of such a facility, the "Parallel Machine Scenario". It was reasoned that such a technology testing facility, in parallel with ITER, could cut 10–15 years off the time it would take to get the data needed for a demonstration power plant (Demo) compared to relying on a second, technology testing, phase of ITER for that data.

Ultimately, the President would propose a $22 million increase for magnetic fusion for FY 1993 (to $360 million) but Congress would only appropriate $327 million. I referred [74] to the events of the 1989–1992 period as the US government's "on-again, off-again fusion strategy". Paul Gilman, an aide to New Mexico Senator Pete V. Domenici, stated [75], "Washington is ruled by one principle: there is no money". Gilman would, in late 1992, join the White House Office of Management and Budget (OMB) as associate director for Energy and Natural Resources, Energy, and Science.

Major Achievement: JET

In late 1991, scientists at the Joint European Torus (JET) in the UK produced 2 million joules of fusion energy over a 2-s pulse, with a peak power of 1.7 MW. This was the first time tritium (which enhances the fusion rate in a pure deuterium plasma almost a hundredfold) was used in a tokamak device (only a 10% tritium mixture was used). About 15 MW of heating power was used in the experiments. Princeton Plasma Physics Director Ron Davidson said, "This is an historic event for fusion". I was quoted in the November 11 New York Times and the November 12 USA Today as saying, "The JET work marks the beginning of the actual use of fusion fuel. This marks the beginning of the transition from research to reality". JET Associate Director Alan Gibson was quoted in the November 20 Christian Science Monitor as saying, "At last we have succeeded. It is a real milestone". JET Director Paul-Henri Rebut was quoted in the November 20 Christian Science Monitor as saying of the JET results, "It confirms Europe's leading position in fusion research. We are ahead of our main rival in the U.S., the Princeton Plasma Physics Laboratory".

Indeed, a friendly competition between JET and the US device TFTR at Princeton had been going on since the two projects were authorized for construction about the same time in the mid-1970s. US scientists had been confident that they would be first to introduce tritium in their experiment until late 1988, when the fiascoes described in the last chapter resulted in a DOE order to Princeton to stop all work on the tritium systems. That action had resulted in the layoff of 160 industry personnel from the TFTR project on 1 day's notice. Work on preparations to use tritium in TFTR was restarted 2 years later, in late 1990, and tritium operations (with 50–50 DT mixture) in TFTR began in December 1993.

ITER Moves Forward

In November 1991, representatives from the USA, European Commission, Japan, and USSR reached accord on an ITER Agreement for review by higher-up officials of their governments. The Agreement, which would later be ratified, launched a

6-year ITER Engineering Design Activity (EDA) phase of the project as a follow-on to the recently concluded Conceptual Design Activity (CDA). Evgeny Velikhov (USSR) was named chair of an ITER Council to oversee the project, with Masaji Yoshikawa (Japan) as cochair and chair of an ITER Management Advisory Committee. JET Director Paul-Henri Rebut was named ITER project director. Paul Rutherford (US) was named chair of an ITER Technical Advisory Committee. Design sites were subsequently established in San Diego, Garching (Germany), and Naka (Japan). Alexander Glass (LLNL) was named US ITER home team leader.

In 1992, Glass formed an ITER Industry Council (IIC), chaired by Harold Forsen (Bechtel Group, Inc.). At its first meeting, March 3, 1992, the IIC "expressed concern that industry would still be perceived as being subservient to the laboratories in the proposed (by Glass) arrangement, and recognized that it would be up to the Home Team Leader to ensure that industry would assume the lead role in design, manufacture, and testing of prototype technology". The IIC "called for a broader role for industry in the whole MFE program and not just in ITER". William Ellis (Ebasco/Raytheon) would later chair the IIC.

Another Fusion Law

On October 7, 1980, President Carter had signed the Magnetic Fusion Energy Engineering Act of 1980 calling for operation of a fusion demonstration power plant by the year 2000. Then, the next month, he lost his bid for reelection to Ronald Reagan. On October 24, 1992, President George H. W. Bush signed the Energy Policy Act of 1992. The next month he lost his bid for reelection to Bill Clinton.

The Energy Policy Act of 1992 directed the Secretary of Energy to "conduct a fusion energy 5-year program that by the year 2010 will result in a technology demonstration which verifies the practicality of commercial electric power production". The Act also stated that "within 180 days after the date of enactment of this Act, the secretary shall prepare a comprehensive management plan for the fusion energy program. This plan shall include specific program objectives, milestones and schedules for technology development, and cost estimates and program resource requirements". The Act authorized $340 million for FY 1993 and $380 million for FY 1994 for fusion. The Congress would later provide $327 million for FY 1993 and $322 million for FY 1994.

In a letter dated October 20, 1992 to Energy Secretary Watkins, the Secretary of Energy Advisory Board (SEAB) commented on the fusion community's proposed plan to build a new Tokamak Physics Experiment (TPX) saying, "we believe that this proposed project is of solid technical merit" and "recommend its further development". DOE then sent a request to OMB for the start of TPX "Title 1 Design" in FY 1994.

In November 1992, Bill Clinton won election over George H. W. Bush to become President-elect of the United States, and, on December 21, 1992, he named Hazel R. O'Leary as Secretary-designate of the Department of Energy. O'Leary was, at the time, executive vice president of Northern States Power in Minnesota.

Northern States Power had been a corporate affiliate of Fusion Power Associates since 1980. During the Carter Administration, O'Leary had been head of the Energy Regulatory Administration, while her late husband, John O'Leary, was undersecretary of the Department of Energy.

Although Secretary Watkins had produced a National Energy Strategy and Congress had just passed the Energy Policy Act of 1992, President-elect Clinton, in designating O'Leary as his choice for Energy Secretary, said that she was "ready to give new life to a department that has failed to meet the pressing national needs it was created to address".

As he took office, President Clinton did endorse an expanded fusion effort. On February 17, 1993, President Clinton spoke to a joint session of Congress outlining his plan to stimulate the US economy and reduce the federal deficit. A White House document accompanying the President's address entitled "A Vision of Change for America" states, "Fusion offers the promise of abundant energy from readily available fuels with low environmental impact. The centerpiece of the research effort in magnetic fusion energy is a collaboration among the United States, the European Community, Japan, and Russia to build an International Thermonuclear Experimental Reactor (ITER). Design and construction of ITER will be a multibillion-dollar effort that would take two decades to complete. The United States must maintain a vital domestic research program to support our efforts on ITER. Yet, the U.S. has not commissioned a major new machine for fusion research since the early 1970s. This investment (which Clinton was proposing) would fund moderate growth in the U.S. fusion energy program above inflation to allow construction of a new facility, the Tokamak Physics Experiment (TPX)".

As President Bush left office in January 1993, he left behind a proposed FY 1994 fusion budget request for \$423 million, up from a proposed \$340 million in FY 1993. The proposed budget contained \$36 million for the initiation of the TPX project. President Clinton subsequently requested only \$348 million, which Congress eventually appropriated.

The 1992 and 1994 EPRI Fusion Reviews

In October 1991, Bob Hirsch, who had refocused the US fusion program on its energy goal in the 1970s and had since held executive posts with both Exxon and Arco, became a vice president of the Electric Power Research Institute (EPRI), in charge of their Washington, DC office. He was hired by Richard Balzhiser, then president of EPRI, who had been influential in fusion policy in the 1970s as an associate director of the White House Office of Science and Technology Policy (OSTP) and who had arranged to fund the budding laser inertial fusion program at the Naval Research Laboratory (Chap. 4).

Hirsch set up the 1992 EPRI Fusion Review Panel. He chaired the panel. Other members were Floyd Culler (a former president of EPRI and former director of Oak Ridge National Laboratory), Nari Hingorani (vice president of the EPRI Electrical Systems Division), John Taylor (vice president, nuclear power, at EPRI), Thomas Schneider (Executive Scientist and comanager of the EPRI Office of Exploratory and Applied Research), and Dwain Spencer (vice president, commercialization and business development, and member of the executive staff at EPRI).

The purposes of the EPRI study were the following:

- To evaluate a wide range of fusion concepts from a utility desirability standpoint
- To enhance EPRI's perspective on fusion
- To provide guidance to DOE on fusion concept characteristics important to utilities
- To provide a basis for reestablishing DOE-EPRI communication and cooperation in fusion

The Panel developed a set of "operational considerations to be used in assessing fusion concepts from the point of view of their desirability to an electric utility". These were:

- Complexity of the reactor configuration
- Availability characteristics of various fusion configurations
- Fuel choice and fuel cycle of various fusion configurations
- Energy balance and subsystem efficiencies of various fusion configurations
- Safety aspects of various fusion configurations
- Waste streams associated with various fusion configurations
- Siting issues associated with various fusion configurations
- Uncertainties associated with the physics/technologies of various fusion configurations

The Panel, in its report [76], concluded that "the federal fusion research program represents an important national investment" and that "producing deuterium–tritium fusion power in the 10–20 MW-thermal range in the Princeton TFTR is an important program milestone and should continue to be a high priority". The panel commented that the tokamak configuration "is highly complex" and said, "program diversity beyond tokamaks is important".

They recommended that, in diversifying its fusion program, DOE should give special consideration to the following:

- Concepts and/or designs that may be less complex
- Power plant designs without tritium burning, because of the very serious materials problems associated with 14 Mev neutrons
- Use of certain low activation materials
- High overall energy conversion efficiency, e.g., combined direct electrical and thermal conversion
- The outage and waste disposal problems of changing out large volumes of fusion reactor core materials every few years
- The importance of effective ash removal from fusion plasmas

The Panel recommended that "engineering thinking and eventual needs of the marketplace should become a critical element in fusion program planning and decision-making".

The 1992 EPRI fusion panel consisted completely of EPRI executives. In 1994, Hirsch assembled a new fusion panel consisting almost exclusively of utility executives. The makeup of the 1994 panel is shown in Table 7.3. The panel issued a report, "Criteria for Practical Fusion Power Systems [77]". The panel listed "three principal types of criteria" of "overarching importance:" (1) economics, (2) public acceptance, and (3) regulatory simplicity.

Table 7.3 Members of the 1994 EPRI Fusion Review Panel

Jack Kaslow, Chair, EPRI	Bill Muston, TU Electric
Merwin Brown, Pacific Gas and Electric Company	Art Peterson, Jr., Niagara Mohawk
Robert Hirsch, EPRI	Steve Rosen, Houston Lighting and Power Company
Ralph Izzo, Public Service Electric and Gas Company	Thomas Schneider, EPRI
John McCann, Consolidated Edison Company of New York	Peter Skrgic, Allegheny Power System, Inc.
Dennis McCloud, Tennessee Valley Authority	Bruce Snow, Rochester Gas & Electric Corporation

With regard to economics, the panel said, "the ability to finance early fusion plants will require a high level of confidence in the performance of a commercial plant. Convincing validation of performance in demonstration or pilot plants will be needed to gain that confidence". They said, "to compensate for the higher economic risks associated with new technologies, fusion plants must have lower life-cycle costs than competing proven technologies available at the time of commercialization".

With regard to public acceptance, the panel said, "a positive public perception can be best achieved by maximizing fusion power's environmental attractiveness, economy of power production, and safety". They listed numerous characteristics that would contribute to environmental attractiveness, including avoiding or minimizing radioactive wastes, toxic emissions, and waste heat.

With regard to regulatory simplicity, the panel said, "plant and systems design will influence regulatory requirements". They indicated the following important design considerations: "avoidance of the need to separate the plant from population centers or for off-site emergency planning; minimizing the need for engineered safety features; minimizing waste production; and minimizing occupational exposure to radiation".

Hirsch had become disaffected by the emphasis on the deuterium–tritium fusion fuel cycle. In a March 5, 1993 presentation to the DOE's Fusion Energy Advisory Committee (FEAC), Hirsch said, "DT tokamak and laser fusion reactors as currently envisaged will be extremely complex, highly radioactive, likely to be highly regulated and costly". He recommended that the fusion program "get off the DT fuel cycle" and "scale up alternate R&D concepts as fast as possible". Harold Forsen (Bechtel Group, Inc.), however, told a House of Representatives hearing on May 5, "I worry very much about not getting on with DT-fueled tokamaks at the expense of too much second guessing of what might be cheaper, smaller or fueled with more exotic, remote fuels". "To say today that any fusion power plant is or is not acceptable is to anticipate a future that will be very different than it is today," he said.

The magnetic fusion community prepared a series of fusion power plant designs during the 1990s, under the direction of Bob Conn and later Farrokh Najmabadi at UCSD. In 1996, an EPRI Fusion Working Group, chaired by Steve Rosen (Houston Lighting and Power), reviewed the latest design called ARIES-RS. In a letter to the DOE Director of Energy Research after their September 17–18, 1996 meeting,

Rosen said, "We were often critical of what we saw (in the past) in that the existing preliminary designs and draft standards then available were not fully responsive to our new criteria". However, Rosen said, "The safety standards have been revised to directly address our criteria and the ARIES-RS conceptual design appears to have carefully considered our inputs in order to provide a useful starting point for further design evolution within the envelope of the newly approved safety standards". He said, "while we would like to see a further reduction in the projected cost of energy, the ARIES-RS conceptual tokamak power plant design has many of the features we, as end-users, find attractive in a future power system".

Inertial Fusion and Beginnings of the NIF

Following the advice of the DOE Fusion Policy Advisory Committee (FPAC), the DOE established an Inertial Confinement Fusion Advisory Committee (ICFAC). The ICFAC held its first meeting on December 16–18, 1992. Following that meeting, the ICFAC sent a letter report to Secretary Watkins recommending that "DOE should proceed with an effort to develop a conceptual design for a National Ignition Facility (NIF)". The goal of the NIF, they said, should be "a cost-effective 1–2 MJ, 500–700 Terawatt advanced glass laser laboratory facility with the goal of demonstrating and studying the regime of ignition and modest gain". The ICFAC letter further stated, "while ignition physics is the most urgent goal, the committee firmly believes that the long range future of ICF will be best served by continuation of other driver programs (light ions, KrF, direct drive) which are likely to be better suited for LMF (Laboratory Microfusion Facility) and/or energy applications". The members of ICFAC are shown in Table 7.4.

Table 7.4 Members of DOE Inertial Fusion Advisory Committee (ICFAC), 1992

Venkatesh Narayanamurti, Chair	Donald Dudziak, North Carolina State University
University of California, Santa Barbara	David Hammer, Cornell University
Solomon J. Buchsbaum, Bell Laboratories	Arthur Kermin, MIT
Timothy Coffey, Naval Research Laboratory	Steven Koonin, California Institute of Technology
J. Richard Airy, SAIC	Gerald Kulcinski, University of Wisconsin
Betsy Ancker-Johnson, World Environmental Center	Conrad Longmire, Mission Research Corporation Bruce Miller, Titan Spectron Corporation
John Birely, US Department of Defense	Marshall Rosenbluth, University of California, San Diego
Robert Christy, California Inst. of Technology	William Simmons, Consultant
Ronald Davidson, Princeton University	Alvin Trivelpiece, Oak Ridge National Laboratory
Anthony DeMaria, United Technologies	

On December 7, 1993, the DOE finally declassified much of the research on inertial confinement fusion, thereby satisfying the 1990 recommendation of the FPAC and subsequent promise to do so by Energy Secretary Watkins. Though the declassification was not as extensive as many would have liked, it was a clear sign of change toward openness and opened the door to more extensive international collaborations. At the December 7 press conference announcing the decision, DOE claimed that now only 20% of ICF would remain classified. Energy Secretary Hazel O'Leary then launched a major review aimed at declassification of other DOE documents dating back many decades.

Signs of Trouble

Senator J. Bennett Johnston, chair of the Senate Appropriations Committee and a supporter of fusion, nevertheless wanted to see the fusion program focus on rapid development of the tokamak. In his view the USA should commit to the construction of ITER and focus the domestic fusion program on ITER support. He was

Major Accomplishment: TFTR

On Thursday night, December 9, 1993 at 11:08 P.M. EST, scientists at the Princeton Plasma Physics Laboratory produced 3 million watts of fusion power using deuterium–tritium fuel in the Tokamak Fusion Test Reactor (TFTR). The next day the power was increased to 6 million watts, exceeding the previous record of 1.7 million watts produced in the Joint European Torus (JET) 2 years earlier. The power was released in a pulse lasting about 1 s. It was the first time a 50–50 mixture of deuterium and tritium had been used in a fusion experiment (the JET result was obtained with only a 10% tritium content). The TFTR experiments reached temperatures of about 350 million degrees, well above the roughly 100 million degrees needed for a fusion reactor. The temperature was achieved using about 20 MW of neutral beam heating power. This accomplishment was especially pleasing to me, since I had prepared the AEC "decision paper" that led to the approval of its construction in 1974. Dr. Richard Balzhiser, president of the Electric Power Research Institute (EPRI), sent a letter dated December 22 to PPPL director Ron Davidson saying, "Your demonstration of over 5 MW of fusion power in the TFTR is a magnificent achievement. We in the utility industry applaud your efforts and this very promising result". The TFTR would later (November 1994) achieve 10 million watts of fusion power (Fig. 7.1), thereby satisfying its original design objective. The TFTR project was under the direction of PPPL scientist Dale M. Meade and later Richard Hawryluk.

TFTR later also demonstrated plasma heating by products of fusion reactions (helium nuclei) for the first time.

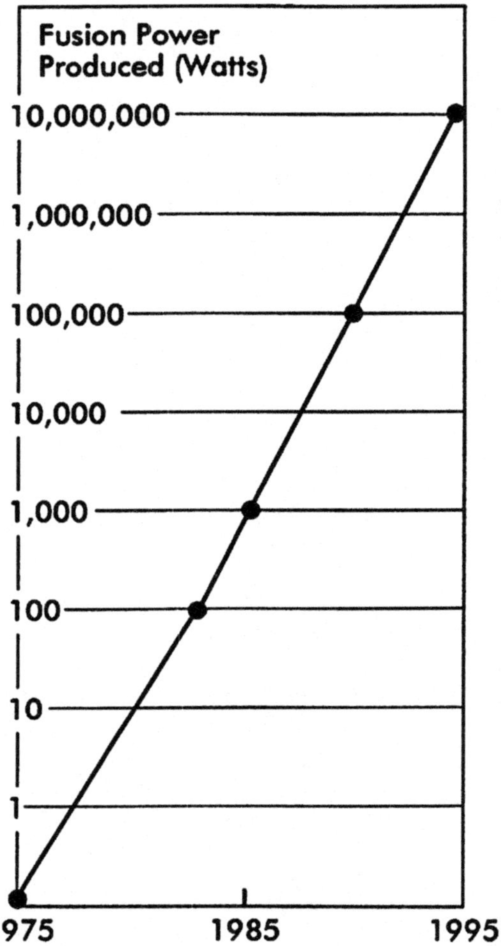

Fig. 7.1 Fusion power produced in tokamak experiments over the period 1975–1995 rose from less than 1 W to over 10 million watts

unhappy that ITER was continuing for so many years as a design effort, with no agreement on siting or construction. At one point in 1993, he introduced a bill renaming the US fusion program to be called the ITER Program. The bill stated that if the USA did not concentrate all its efforts on rapid accomplishment of ITER, the program budget would be reduced to $50 million per year. The bill did not pass. However, on June 23, 1994, as part of the FY 1995 appropriations process, the Senate Committee cut President Clinton's fusion request by $10 million and limited TPX to design only. In a report accompanying the Senate mark, the Committee stated that it was "very concerned that the Department has failed to report to Congress on how it intends to move forward with the International Thermonuclear Experimental Reactor (ITER) and to address the complex domestic and international issues associated with a project of this magnitude. The Department was directed last year to provide a plan that would describe the selection process for a candidate site within the United States for ITER, the necessary steps that will lead to the final selection process of a host site for ITER and the schedule and critical path including milestones and budgets that would be necessary for the design, construction, and beginning of operation of ITER by 2005. These plans are essential to assure an orderly transition from the beginning of engineering design of ITER to the beginning of construction, now scheduled for 1998. The Administration's reluctance to put forward a framework jeopardizes the future of the US fusion program". The report adds, "without ITER, TPX's contribution to the Department's fusion program would be questionable". The report states, "We strongly believe we should not pursue TPX unless and until both the President and the Congress have made a full commitment to ITER".

One week later, on June 30, 1994, the full Senate voted 60–30 to restore the cut and authorize the TPX. Two weeks later, on July 13, President Clinton wrote a letter to New Jersey Governor Christine Todd Whitman stating, "I am committed to pursuing a strong, balanced program for the development of fusion energy, and I'm encouraged by your support for this promising area of science". Clinton said, "The TPX project at Princeton is the next major step for the US fusion program, and I am confident that it will prove fusion technology to be a safe and commercially attractive form of power for the future. My budget for fiscal 1995 includes funding for TPX". "I look forward to working with you in the future to help our nation realize the great possibilities of fusion power". Soon thereafter House and Senate passed the FY appropriation bill, providing the full amount for fusion requested by the President, but limiting TPX to design only. They also called for a review of fusion policy by the President's Council on Science and Technology (PCAST).

In October 1994, Energy Secretary Hazel O'Leary formed a task force on Strategic Energy Research and Development of the Secretary of Energy Advisory Board. I was 1 of 31 persons appointed to the panel, which was chaired by Daniel Yergin (Cambridge Energy Research Associates). O'Leary said she was calling for "a high level review of the Department's \$1.8 billion portfolio of applied energy programs". She said she wanted "to get an independent assessment of whether the Department has a coherent and effective plan for getting energy technologies into the marketplace". At its first meeting, October 12, O'Leary told the task force that funding for energy programs had declined fivefold since 1978, from \$9.2 billion to \$1.8 billion (1995 dollars), even though world energy use had nearly doubled during this period and US oil imports were currently running over \$40 billion per year. The task force was set up to develop the case for increased funding for DOE energy programs. They were soon to be diverted from that task, however.

8 Financial Tsunami: 1995–1999

A Molehill Can Be a Mountain to a Sparrow

R. Godden
Author, An Episode of Sparrows

Contract with America

Congressional midterm elections were held in the USA in November 1994. As a result, the Republican Party took control of both the House of Representatives and the Senate in January 1995. When control of either Congress or the Presidency passes from Democrats to Republicans (or vice versa), the inevitable result is that, regardless of the merits, whatever was being done previously is discounted and "change" is required. Rep. Newt Gingrich was elected Speaker of the House. He announced a new policy to be called "the Contract with America." In practice, this meant "cut federal spending." The Contract with America called for cutting the fusion program by about 50% [78].

President Clinton (a Democrat) was in the process of finalizing his FY 1996 budget submission to Congress. At a meeting of the DOE Energy R&D Task Force, of which I was a member, Energy Secretary O'Leary told our panel that she had been told by the White House to plan on a $1.2 billion reduction at DOE. She changed the charge to the Task Force from one of justifying increased spending on energy programs to one of advising her on how to carry out viable energy programs with reduced budgets.

Despite this negative budget climate, President Clinton sent a FY 1996 budget request to Congress that included funds for construction of the TPX magnetic fusion experiment at PPPL and the construction of a new inertial fusion experiment, the National Ignition Facility (NIF) at LLNL. He requested $366 million, compared to $365 million (FY 1995), for magnetic fusion and $241 million, compared to $177 million, (FY 1995) for inertial fusion.

S.O. Dean, *Search for the Ultimate Energy Source: A History of the U.S. Fusion Energy Program*, Green Energy and Technology, DOI 10.1007/978-1-4614-6037-4_8,

The National Ignition Facility

The National Ignition Facility (NIF) facility, located at the Lawrence Livermore National Laboratory (LLNL) in California, is aimed at using an array of laser beams (192) to ignite fusion reactions in small capsules containing deuterium–tritium fuel. The facility, as proposed in the FY 1996 budget, would use approximately 1.8 MJ of laser energy to produce the same or a little more amount of fusion energy. It was justified as a vehicle to study weapons physics in the absence of underground nuclear weapons testing and as a stepping stone to the previously considered Laboratory Microfusion Facility (LMF) that had the goal of producing up to 100 times more fusion energy than the input laser energy. The cost of the NIF was estimated at $1.1 billion, with operation targeted for 2002. NIF would ultimately cost approximately twice that estimate and operate in 2009.

NIF lasers are single-shot, whereas a power plant would require lasers (or other types of drivers) that fire repetitively, perhaps ten times a second. DOE named David Crandall to be Director of a new Office of the National Ignition Facility. Dave had previously held the position of Director of the Advanced Physics and Technology Division in the Office of Fusion Energy.

In early 1995, the DOE held a series of public comment workshops on NIF. Three public workshops were held (January 24, January 30, and March 9) on NIF and its potential relation to nuclear weapons proliferation. Dozens of persons attended to voice their opposition to NIF construction, arguing that a facility to study weapons physics, that would also be available for international collaborations, would lead to nuclear weapons proliferation. Some of the speakers also argued that NIF construction was just a subterfuge for keeping up employment levels at the DOE weapons laboratories; others were opposed because the increased expenditures went against the current budget climate in Washington to reduce federal expenditures in general. Two of the most aggressive NIF opponent groups, which would keep up their opposition throughout NIF construction, were the National Resources Defense Council (NRDC) and Tri-Valley CAREs, a Livermore Valley antinuclear civic group.

I spoke at the March 9 workshop. I said, "NIF is a timely and necessary experiment to establish the scientific principles of inertial confinement fusion, a technology that has numerous commercial benefits, including eventually the development of a commercially competitive energy source." I noted that the basic principles of nuclear weapons design have been widely known to all nations for a long time and do not require the physics that will be established in NIF. I said, "Proliferation policy should be based on the control and inspection of fissionable material and related equipment, not based on the suppression of scientific investigations of the fusion process." I noted that the JASONS, a prestigious scientific group, had reviewed the NIF from the proliferation standpoint and had concluded "the NIF is an extremely sophisticated challenge, not one which could conceivably be undertaken by, or useful to, a potential proliferator."

The 1995 PCAST Fusion Review

The FY 1995 appropriation bill asked the Executive Branch to carry out a review of the fusion program by the President's Council of Advisors on Science and Technology (PCAST). The review began in March 1995. Professor John P. Holdren (University of California, Berkeley), a PCAST member, chaired the review. Holdren had previously made several important contributions to the fusion program, including a study he led on the environmental, safety, and economic aspects of fusion [79]. In setting up the study, Presidential Science Advisor Jack Gibbons said, "In the twenty-first century, increasing attention must be devoted to developing energy sources that can supply large quantities of electricity in an environmentally sustainable manner. It is within this context that the role of fusion and other energy supply technologies should be evaluated."

The panel was charged "to conduct a review that identifies the technical and policy tradeoffs and budgetary requirements for at least four different options for structuring the magnetic fusion program." The charge said, "The committee should proceed from the assumption that eliminating the fusion program is not an option under consideration."

The four options provided to the committee were as follows:

- Build TPX and join the next phase of ITER.
- Build TPX and do not join the next phase of ITER.
- Do not build TPX but join the next phase of ITER.
- Do not build TPX and do not join the next phase of ITER.

The US fusion community viewed construction of TPX (Tokamak Physics Experiment) as the centerpiece of the US domestic magnetic fusion program, aimed at studying steady-state tokamak physics. The TPX, though located at Princeton Plasma Physics Laboratory, was a national project, overseen by a TPX National Council. Stewart Prager (University of Wisconsin) chaired the TPX National Council. I was on the Council, whose other members included David Baldwin (LLNL), Ronald Davidson (PPPL), John Dawson (UCLA), Gerald Navratil (Columbia University), Miklos Porkolab (MIT), Paul Rutherford (PPPL), John Sheffield (ORNL), and Richard Siemon (LANL).

The charge to the PCAST committee also said, "the committee may consider other options if it chooses to do so." The charge told the committee to take into consideration the DOE's tight financial situation, including an announced plan to cut overall energy funding over the next 5 years, and to take into account the ongoing activities of the DOE's Energy R&D Task Force (of which I was also a member), "which has been charged with reviewing the Department's portfolio of applied energy R&D programs." In addition to chairman Holdren, the other members of the PCAST panel were Norman Augustine (Lockheed Martin), Robert Conn (University of California, San Diego), Lawrence Papay (Bechtel), Andrew Sessler (Lawrence Berkeley National Laboratory), Robert Sokolow (Princeton University), Charles Vest (MIT), and Lillian Wu (IBM). At their invitation, I met privately with the PCAST panel on April 24, 1995.

The PCAST committee issued its report on June 16, 1995. It noted that the combined fusion efforts of Europe and Japan "already total more than three times the corresponding effort here." They stated, "we believe there is a strong case for the funding levels for fusion currently being proposed by the U.S. Department of Energy—increasing from $366 million in FY 1996 to about $860 million in FY 2002 and averaging $645 million between FY 1995 and FY 2005." They acknowledged, however, that "Although the program just described is reasonable and desirable, it does not appear to be realistic in the current climate of budgetary constraints." In light of these constraints, they concluded, "We urge, therefore, that the Administration and the Congress commit themselves firmly to a U.S. fusion R&D program that is stable at not less than $320 million per year." This level was $46 million per year less than the level at the time. The committee said that TPX could not move forward in the $320 million case. Instead, they urged US officials to ask the international ITER team to downsize ITER to a much less ambitious "burning plasma experiment" similar to the previously US-proposed BPX. The committee said that if the program were reduced to about $200 million per year, it "would leave room for nothing beyond a core program of theory and medium-scale experiments."

At a July 26–27, 1995 ITER Council meeting, the ITER Council firmly rejected the PCAST recommendation that the technical objectives of ITER be reduced and refocused on the design of an ignition physics test facility at one-third the estimated cost of ITER. The Council stated, "There is a renewed consensus of the ITER Council that this is a necessary step; that the objectives of ITER remain attainable and must not be changed; that the design can meet the objectives; that the quadripartite cooperation has shown to be an efficient frame; and that the right time for such a step is now." The members of the ITER Council were Evgeny Velikhov (Russia), chair; Paolo Fasella (EEC); Naotaka Oki (Japan); and James Decker (USA). Decker was Acting Director of the DOE Office of Energy Research at the time.

At a joint meeting of Fusion Power Associates and the Canadian Nuclear Association in Montreal, September 7, 1995, ITER Council Chairman Velikhov suggested that consideration be given to bringing in other partners in order to raise the funds necessary for ITER construction. He mentioned Korea, India, and China as possibilities. He said it was logical for Asian nations to take a more active role in developing fusion "since that is where the future energy crisis will be most acute." On July 23, 1995, Korean President Kim Young-sam had announced an ambitious plan to join the world effort to develop fusion as an energy source as part of a greatly expanded effort in Korea to develop energy, space, and other advanced technologies. Kim said that Korea would start designing its fusion facilities immediately and start building them in 1998.

Congress Takes Aim

Although TFTR had only recently achieved its original design objective of producing over 10 MW of fusion power, and President Clinton had protected the fusion budget in his otherwise stringent FY 1996 budget request, Congress was more focused on making massive cuts in federal programs, including fusion.

On March 16, 1995, the House Budget Committee voted to direct DOE to begin termination of the entire agency in FY 1996, as part of a proposed $190 billion (over 5 years) federal spending reduction package. In an appended report, the committee listed possible programs to eliminate, including "the international fusion program." At a February 15 hearing, Rep Dana Rohrabacher, chair of the House Science Committee's Subcommittee on Energy and Environment, stated, "We must decide if the modest success shown (in fusion) for the billions spent is worth billions more required to continue the program for at least another 30 years." At a March 14 hearing of the Senate Appropriations Committee, Chairman Pete V. Domenici (a long-time fusion supporter) said, "if we go with fusion, it will go up and up and up and we're not going to have money for other things."

Several members of the fusion community were invited to testify at a March 28 hearing of the House Appropriations Subcommittee on Energy and Water, which oversees the Department of Energy's budget. Those testifying included Bob Conn (University of California at San Diego and a member of the PCAST fusion review panel), Ron Davidson (Director of PPPL), Jim Drake (University of Maryland, on behalf of the University Fusion Association), Bruce Montgomery (MIT), Stewart Prager (University of Wisconsin, on behalf of the American Physical Society Division of Plasma Physics), Ned Sauthoff (PPPL, on behalf of the Institute of Electrical and Electronic Engineers), and myself.

I traced the history of the fusion program over the previous 20 years and stated that "there has been progress commensurate with the funding provided." I stated, "Fusion and other advanced energy technologies are absolutely necessary for the survival of advanced industrial civilization. The money invested in fusion now is a modest and prudent investment to ensure a high standard of living for future generations."

Davidson stated, "By any measure, the world must find new sources of energy in the coming decades." He said, "Well before the middle of the next century, the world faces an energy deficit of extraordinary proportions." He noted that the Tokamak Fusion Test Reactor (TFTR) at Princeton had produced "world-record levels of up to 10.7 million watts of fusion power in November 1994," and, he said, "with these historic experiments, I believe that the development of fusion energy has moved into a new era."

Conn said, "Short of building new machines, it's hard to show you more dramatic results than what the program recently produced." He remarked, "We're ready now to start construction on a new U.S. tokamak which will enable further optimization of the fusion power plant design."

Drake noted that "the U.S. presently imports about 50% of our yearly oil consumption at a cost of about $100 billion" and commented, "the annual cost of the fusion program is less than 0.4% of the annual cost of our oil import bill." He said, "the investment in this program (fusion) is dwarfed by the potential benefits."

Prager noted that "the fusion program has given birth to a new branch of science—plasma physics" which "produces results of deep scientific value, and immediate and vast application." He indicated that "the annual expenditure on fusion is less than one-thousandth of the annual expenditure on energy." He said that the fusion budget "has been about halved in the past decade."

Sauthoff said, "The fusion program should include an appropriate balance of ignited plasma studies, such as in TFTR and ITER, fusion technology programs, tokamak concept improvements as in the current base program and the planned TPX, inertial fusion energy such as in the National Ignition Facility, alternate concepts and basic plasma studies."

Shortly before the issuance of the June PCAST fusion review committee report described above, the House Science Committee and the House Appropriations Committee marked up the FY 1996 fusion budget at $229 million. House and Senate appropriators eventually agreed to provide $244 million, a 33% reduction of $119 million from the FY 1995 level of $363 million. The appropriation provided no funds for construction of the TPX.

Although the US fusion program had experienced budget reductions in the past, a 33% reduction in 1 year was unprecedented and resulted in wholesale layoffs, program terminations, and delays in remaining programs. According to a press release from the Princeton Plasma Physics Laboratory, the largest of the fusion laboratory programs, they were issuing "layoff notices to 166 regular employees and 80 personnel hired through subcontractors" out of a total pre-layoff total of 820 regular employees and 110 subcontractor staff. DOE told PPPL to stop operation of their major experiment (the TFTR). In addition, DOE ordered the shutdown of the TEXT tokamak at the University of Texas and the Phaedrus experiment at the University of Wisconsin. DOE also indicated that it might have to cease operation for a year or more on the other two major US tokamaks (at MIT and General Atomics) and scale back its participation in the ITER project.

Both House and Senate appropriations bills called for a "restructuring" of the magnetic fusion program to meet lower projected funding levels. The report accompanying the final appropriation bill stated, "With little prospect for increased funding for the fusion program over the next several years, it will be necessary for the program to restructure its strategy, content, and near- to medium-term objectives. The restructured program should emphasize continued development of fusion science, increased attention to concept improvement and alternative approaches to fusion, and development and testing of the low activation materials so important for fusion's attractiveness as an energy source." DOE charged its Fusion Energy Advisory Committee (FEAC) to make recommendations on this restructuring with a report due in December 1995.

Congress did not cut the inertial fusion program, which is funded within DOE's weapons category, approving $241 million compared to the FY 1995 level of $173 million. Funds for proceeding with engineering design and construction of the National Ignition Facility were provided in the appropriation.

Industry Groups Respond

At a joint meeting of the Fusion Power Associates Board of Directors, the Fusion Industry Council US (FICUS), the US ITER Industry Council (IIC), and the TPX Industry Council on September 13, 1995, it was decided to write a letter to Energy

Secretary Hazel O'Leary regarding developing a new fusion strategy. The letter stated, "In adjusting to whatever budget is eventually passed, we ask that you pay special attention to assuring that a balanced effort, which provides for both scientific advances and technology development leading eventually to a commercial energy source is maintained." The letter stated, "Fusion requires continued evolution of three synergistic elements: new science, development of relevant technologies, and a systems approach to program management to optimize the use of limited resources and minimize the time to commercial application." The letter concludes, "Industry will be the means by which the benefits of fusion energy are brought to society, and industry thus has an interest in keeping the fusion energy option open. We believe the private sector has much to offer in a balanced fusion program at any budget level."

The letter was signed by David Baldwin (General Atomics), S. Locke Bogart (Lockheed Martin), James Conner (Babcock &Wilcox), Stephen Dean (Fusion Power Associates), William Ellis (Raytheon Engineers and Constructors), David Everson (Everson Electric), Samuel Harkness (Westinghouse), John Landis (Stone & Webster), James Lang (McDonnell Douglas Aerospace East), Chester Lob (Communications and Power Industries), Michael Monsler (W.J. Schafer Associates), William Robinette (TRW), Alan Todd (Northrop Grumman), and Stephen Toth (CBI Industries).

Fusion Energy Mission Abandoned

In response to the Congressional budget cuts, the DOE Fusion Energy Advisory Committee (FEAC) set up two subcommittees: a standing Scientific Issues Subcommittee, chaired by Jim Callen (University of Wisconsin), and an ad hoc strategic planning subcommittee to make recommendations for a restructured fusion program, chaired by Mike Knotec (Battelle Pacific Northwest Laboratory). The members of Knotec's subcommittee are shown in Table 8.1. The Knotec panel was told by DOE to consider only two budget levels for the fusion program: an upper level of $275 million and a lower level of $200 million. The Knotec subcommittee held two meetings in December 1995 (at PPPL and General Atomics) and made its report to the full FEAC (chaired by Robert Conn) at its meeting January 18–19, 1996 in Washington, DC [80].

In September 1990, the Secretary of Energy's Fusion Policy Advisory Committee (FPAC) had recommended that:

- The United States should commit to fusion as a potential energy source.
- The United States fusion program must become energy oriented, with specific goals of an operating demonstration power plant in 2025 and an operating commercial power plant by 2040.
- The United States should take maximum advantage of international collaboration while maintaining a sound domestic program.
- Involvement of the private sector should be an integral part of the energy-oriented fusion program.

Table 8.1 Members of DOE Fusion Energy Advisory Committee (FEAC) 1994

Robert W. Conn, Chair, University of California, San Diego	George R. Jasny, Martin Marietta Energy Systems (retired)
John F. Clark, Battelle Pacific NW Laboratories	Michael L. Knotek, Battelle Pacific NW Laboratories
Thomas B. Cochran, Natural Resources Defense Council	John W. Landis, Stone & Webster Corporation Stephen L. Rosen, Houston Lighting & Power Co.
Harold K. Forsen, Bechtel Hanford Inc.	Marshall N. Rosenbluth, Univ. of California, San Diego
Joseph G. Gavin, Jr., Grumman Corp. (retired)	P. Floyd Thomas, Jr., Martin Marietta Energy Systems
Katharine B. Gebbie, National Inst. of Standards and Technology	James R. Thompson, Jr., Orbital Sciences Corporation
Beverly K. Hartline, Continuous Beam Accelerator Facility	Demetrius D. Venable, Hampton University

In contrast, in early 1996, the FEAC, based on the Knotec panel report, recommended the following policy goals for the restructured fusion program:

- Advance plasma science in pursuit of national science and technology goals.
- Develop fusion science, technology, and plasma confinement innovations as the central theme of the domestic program.
- Pursue fusion energy science and technology as a partner in the international effort.

The Knotec subcommittee (and FEAC) recommended a new "mission statement" for the fusion program: "Advance plasma science, fusion science, and fusion technology—the knowledge base needed for an economically and environmentally attractive fusion energy source."

In 1990, the FPAC had recommended that the fusion program set 2025 as a target date for operation of a fusion demonstration power plant. The restructured program would have no timetable. To the present time (2012), the DOE still has no timetable for operation of a fusion demonstration power plant.

The Knotec panel recommended that DOE change the name of the program from Fusion Energy Program to Fusion Energy Sciences Program and that FEAC change its name to FESAC (Fusion Energy Sciences Advisory Committee). The panel recommended downsizing and limiting the management authority of the DOE Office of Fusion Energy and, instead, use the new FEAC science subcommittee and community peer review for making program decisions, a process they said was more in keeping with the new "science-dominated mission."

A key issue for the Knotec panel was how US participation in ITER could fit into the restructured program. The other ITER partners (Russia, European Commission, and Japan) viewed ITER as a fusion engineering test reactor stepping stone facility to a demonstration fusion power plant. The Knotec panel recommended, "The primary role of the ITER program in the restructured U.S. fusion program is to pursue the science of burning plasmas."

The Knotec panel's recommendations were endorsed by the full FEAC by a vote of 10-2. In a strongly worded dissent, FEAC member Joe Gavin (retired president/CEO of Grumman Corporation and leader of the NASA Lunar Landing Module Project) said that DOE's decision to limit the FEAC considerations to $275 million "represents fundamentally flawed policy." He noted that the President's Council of Advisors on Science and Technology (PCAST) had only recently recommended a level of $320 million. Gavin recommended that FEAC make "a forthright challenge to the Secretary (of Energy) and to the Administration to undertake a determined effort to convince Congress of the merit of providing annual funding for fusion at the PCAST $320 million." Gavin said, "The fusion program in the United States has produced too many successes and has too much promise to give up without a fight, without serious review of national priorities. It will require leadership, not poll-following." The other FEAC member who dissented was J. R. Thompson (former Deputy Administrator of NASA and then executive vice president and general manager, Launch Systems Group, Orbital Sciences Corporation).

Although FEAC chairman Robert Conn (Dean of Engineering, University of California, San Diego) was also a member of the PCAST panel that urged a more ambitious fusion effort, he endorsed the Knotec subcommittee's report and transmitted the report to DOE Director of Energy Research Martha Krebs on January 27, 1996. In his letter of transmittal, Conn said, "The historically strong United States leadership role in the world magnetic fusion energy program came to an end with the decision on FY 1996 funding. However, we conclude that the United States can still play an important supporting role in magnetic fusion energy development, but only by recognizing the new dependence of US efforts on the activities and decisions of Europe, Japan and the Russian Federation. As such, progress will depend on maintaining a balance of domestic and international activities."

At a March 7, 1996 hearing of the House Science Committee Subcommittee on Energy and Environment, Conn testified, "The U.S. program is only about 20% of the world fusion energy effort, making us close to a marginal player in the world scene." He compared the $275 million his committee was recommending to Japan, which he said was spending approximately $450 million per year on fusion, and to the European Community, which he said was spending $600 million. At the same hearing, Joe Gavin criticized the government for its lack of vision and perspective and called for a higher level of fusion spending. "A $100 million increase in fusion spending would amount to only one-fourteenth of the cost of a single B-2 bomber," he said. DOE Director of Energy Research Martha Krebs told the committee that the Department was changing its fusion program "from a goal-oriented energy technology program to a fusion energy sciences program." She said the Department "no longer had any plans for large construction projects." She said that the Department would "remain an active participant in the International Thermonuclear Experimental Reactor (ITER) Engineering Design Activities through its completion in July 1998" but that "given the financial limitations (of the U.S. program) the Department will not seek to be the host of the ITER facility." She indicated that the FEAC will be renamed the FESAC and "its membership changed to reflect the scientific nature of the program." DOE removed almost all the industry personnel from the new FESAC. The membership of the new FESAC is shown in Table 8.2.

Table 8.2 Members of DOE Fusion Energy Sciences Advisory Committee (FESAC) 1996

John Sheffield, ORNL Chair	Mike Knotek, Battell Northwest Pacific Laboratory
Ira Bernstein, Yale University	John Lindl, LLNL
Richard Briggs, SAIC	Earl Marmar, MIT
Jim Callen, University of Wisconsin	Bruce Montgomery, MIT
Patrick Colestock, Fermilab	Marshall Rosenbluth, UCSD
Melissa Cray, LANL	Tony Taylor, General Atomics
Fred Dylla, Thomas Jefferson Accelerator Laboratory	Nermin Uckan, ORNL
Katherine Gebbie, National Institute of Science and Technology (NIST)	Stewart Zweben, PPPL
Richard Hazeltine, University of Texas	Stewart Prager, American Physical Society
Joseph Johnson, Florida A&M University	John Davis, American Nuclear Society
Charles Kennel, UCLA	Ned Sauthoff, IEEE

Writing in the January 1996 issue of APS News, a publication of the American Physical Society, Krebs said, "Personally, I think the Congressional action was unwise, foolish and tragic in the face of what we know will be the energy requirements of the U.S. and the world by the middle of the next century. It is also a tragedy for many individuals who have had a profound commitment to making fusion energy happen. A cut of this size, a shift of direction this sudden, will leave human and scientific wreckage; there is no avoiding it. In spite of this, Congress has made a clear statement and its FY 1996 funding level is based on the expectation that the restructured fusion science program will cost significantly less in the future. This is not the time for denial, delay or recrimination. It is a time for imagination." Krebs said the fusion program must be restructured away from a time-driven effort.

I was no longer a member of FEAC at the time these restructuring events took place. If I had still been a FEAC member, I would have strongly backed Gavin's sentiments. In the June 1996 Fusion Power Associates Executive News letter, I quoted author Thomas W. McKnight, "The difference between a goal and a wish is that a goal has a definite time frame attached to it, within which we intend to do certain things to help accomplish that goal." The USA no longer had a goal to develop fusion as an energy source; it had a wish.

Dr. Samuel D. Harkness, Director, R&D Operations, Westinghouse Research Laboratory, and chairman of the Fusion Industry Council US (FICUS) wrote DOE fusion head Dr. N. Anne Davies July 2, 1996, saying, "Fusion still is, in fact, an energy program focusing on an eventual product rather than a research program solely to advance plasma and fusion science. In this connection, we (FICUS) feel that an emphasis on plasma and fusion science to the exclusion of energy will lead to a continuing downward spiral of congressional support."

Although the magnetic fusion program was being decimated, both financially and strategically, the inertial fusion program was prospering as a weapons-supported enterprise (Congress's budget-cutting zeal did not extend to military spending). When Congress announced the Contract with America to cut federal expenditures, they did not mean it to apply to those programs associated with the military. The inertial confinement fusion program at DOE was within the DOE's nuclear weapons category and was justified as a way to study nuclear weapons physics as part of a newly named "Science-Based Stockpile Stewardship" program. At the same time Congress was cutting the magnetic fusion program funding from $363 million to $230 million between FY 1995 and FY 1997, it increased inertial fusion from $173 million to $367 million. At the same time the magnetic fusion program was prevented from constructing the TPX and being ordered to begin shutdown of TFTR, the Congress authorized construction of the NIF, then estimated to cost $1.1 billion.

Major Advances: Tore Supra, Omega, Z, JET

In early 1996, scientists working on the French superconducting tokamak Tore Supra sustained the plasma discharge for two minutes, a new world record. Tore Supra is the world's first tokamak with superconducting magnet coils. In addition, 90% of the plasma current was driven by 1.9 MW of radiofrequency waves at the lower hybrid frequency.

Also in 1996, scientists at the University of Rochester produced 400 J of fusion energy using 40 kJ of energy from the Omega laser.

In 1997, scientists at the Sandia National Laboratories, using the Z facility, produced 500 kJ of X-rays in a 4-ns pulse (85 TW of power) by driving 7 million amperes of current through a cylindrical array of fine tungsten wires, creating an imploding cylindrical pinch. Large bursts of X-rays are important for driving symmetric implosions of small capsules of fusion fuel in inertial confinement fusion. One year later, they upped their achievement to 2 million joules and 200 TW.

In the fall of 1997, JET carried out a series of ~100 DT pulses reaching fusion power levels of 16 MW for ~1 s and ~22 MJ of fusion energy per pulse using longer-duration lower-power pulses (Fig. 8.1). The maximum fusion gains achieved were $Q = P\text{fusion}/P\text{heat} \approx 0.65$. JET also extended alpha heating experiments and ICRF heating scenarios in DT. A near ITER scale closed cycle tritium plant was tested successfully during this phase. JET also made a major contribution to fusion technology by demonstrating remote handling of components inside the vacuum vessel.

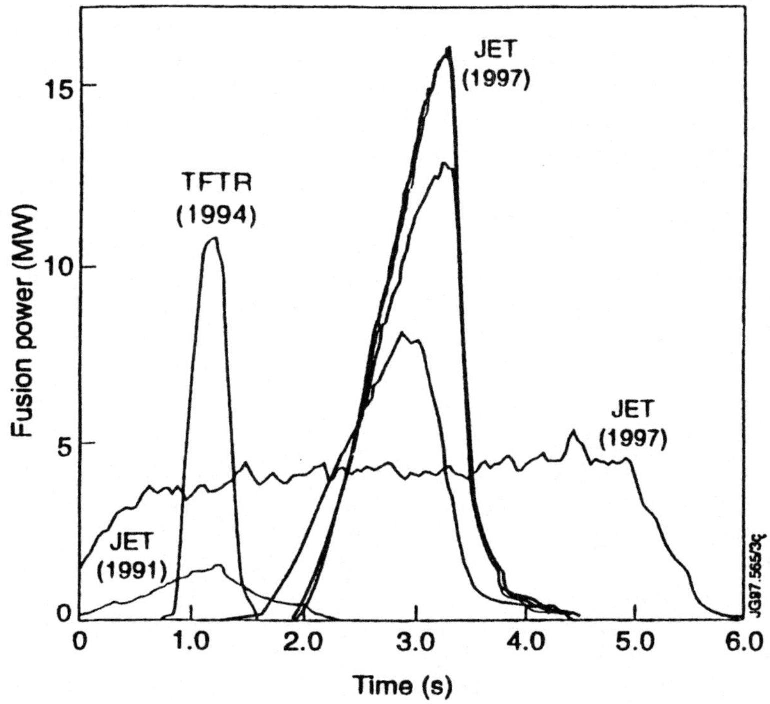

Fig. 8.1 Fusion powers produced in TFTR and JET. Following the achievement of producing 10 MW of fusion power in TFTR in 1994, JET achieved over 15 MW in 1997. JET also produced about 5 MW for about 5 s in 1997 (courtesy of Dale M. Meade, PPPL)

TPX Reborn as KSTAR

The DOE had been unsuccessful in getting the Congress to fund the Tokamak Physics Experiment (TPX), but TPX would soon be reborn and renamed elsewhere. On June 14, 1996, Korean Minister of Science and Technology Dr. KunMo Chung (we had been graduate students together at MIT in the early 1960s) signed an agreement with Energy Secretary Hazel O'Leary to collaborate on fusion energy. Chung announced that Korea and Princeton Plasma Physics Laboratory would collaborate on the design and construction of a major superconducting tokamak in Korea, similar to the ill-fated TPX. Korea would provide funds to Princeton to assist with the design. The device, which was built and is now in operation, was named KSTAR. Many of the activities that the US fusion community had previously identified and advocated as a part of a US program are currently being pursued in Korea and other countries.

In May 1998, Japan began operation of a new billion-dollar-class stellarator, the Large Helical Device (LHD) at the National Institute for Fusion Science, "to develop a scientific basis for a stellarator fusion reactor." Osamu Motojima was head of the LHD experiment. He would later become director-general of the ITER project.

1997 PCAST Energy Report

President Bill Clinton issued a letter dated January 16, 1997 to his science advisor Jack Gibbons "to review the current national energy R&D portfolio, and make recommendations to me by October 1, 1997 on how to ensure that the United States has a program that addresses its energy and environmental needs for the next century." Clinton said, "The analysis should be done in a global context, and the review should address both near- and long-term national needs including renewable and advanced fission and fusion energy supply options, and energy end-use efficiency." PCAST member John P. Holdren (previously a professor at the University of California, Berkeley, and having moved to Harvard) was named to head the study. A nuclear subpanel, chaired by John Ahearne (Duke University), dealt with fusion.

In its report, dated September 30, 1997 [81], "Federal Energy Research and Development for the Challenges of the Twenty-first Century," the new PCAST panel referenced a previous PCAST fusion panel [82] Holdren had chaired in 1995. The new report recommended across the board funding increases for most energy programs, including fusion. The PCAST said these increases were necessary "to close the gap between the current energy R&D program and the one that the challenges require." For fusion, the panel recommended gradual increases from the present $232 million level to a level of $328 million in 2003. They said, "Our Panel reaffirms support also for the specific elements of the 1995 PCAST recommendation that the (fusion) program's budget-constrained strategy be around three key principles: (1) a strong domestic core program in plasma physics and fusion technology; (2) a collaboratively funded international experiment focused on the key next-step scientific issue of ignition and moderately sustained burn; and (3) participation in an international program to develop practical low-activation materials for fusion energy systems."

The PCAST report said, "The present funding level of $230 million is too low in the view of the PCAST Energy R&D Panel; it allows no significant U.S. activity relating to participation in an international program to develop practical low-activation materials; reduces the level of funding for design of the International Thermonuclear Experimental Reactor (ITER); forced an early shutdown for the largest U.S. fusion experiment (TFTR); and canceled the next major U.S. plasma science and fusion experiment. It also limited resources available to explore alternative fusion concepts."

In a letter dated October 3, 1997, the presidents of 16 universities told Energy Secretary Federico Peña "Fusion energy sciences addresses intellectual challenges of fundamental importance and develops applications with clear societal benefits. Our universities are committing academic resources to the excellence of research and teaching in this area." They said, "We are concerned that (further) reductions would endanger the intellectual vitality of this area of science and are writing to express support for Fusion Energy Sciences." The universities represented were Auburn, Columbia, Hampton, MIT, William and Mary, Cornell, Lehigh, NYU, Princeton, Colorado, Maryland, Texas, Wisconsin, and the University of California at Irvine, Los Angeles, and San Diego.

Pathways to Fusion Power Symposium

Fusion Power Associates hosted a 3-day symposium, Pathways to Fusion Power, August 27–29, 1997, attended by 65 participants [83]. Summaries were prepared for the seven topics discussed.

The first topic discussed was "What is our view of the future marketplace?"

The symposium summary noted that the marketplace for electric power in the USA is expected to become increasingly competitive in the near- and mid-term and that current conceptual designs of fusion power plants project costs of electricity higher than today's sources and hence could not compete today on economic grounds. However, the summary says, "In the long-term, projections are difficult. Factors like global warming or regional availability of fuels favor technologies like fusion." The summary says, "Although the opportunity for fusion to compete commercially is decades into the future, it is not too early for the fusion community to seek closer ties with its future customers. At a minimum, their interest is required for political support. However, their technical experience would also be valuable for guiding our R&D program." The summary comments, "It is also important to note that market forces vary around the world, and in many other countries the power industry is more government-controlled than for the US. A favorable trend in the US is the emergence of large nuclear power plant operating companies. These companies may be more receptive to fusion."

The second topic discussed was "Will fusion enter the marketplace as an electric power producer or as a supplier of some other needed product?"

The summary notes, "There are a range of possible commercial and military applications for both fusion itself and for technologies developed for fusion. Some of these might provide early demonstrations of the utility of fusion and/or fusion technologies. It is important to seek out such applications in the near- and mid-term, in view of the economics of the electric power market. Here again, it is important that there be early dialogue with potential customers, for the reasons given above for the electric power market."

The third topic discussed was "Can other fuel cycles compete with the DT fuel cycle?"

The summary says, "Material presented at the symposium indicates that, on a physics basis, it will be difficult for other fuel cycles to compete with the DT cycle. And, as indicated above, even the DT cycle will have difficulty competing with other technologies on economic grounds. Nevertheless, advanced fuels may possess engineering, safety and environmental advantages and hence merit an active ongoing research effort, emphasizing concepts that might be uniquely matched to such cycles."

The fourth topic discussed concerned the tokamak path to commercial fusion.

The summary says, "The development of the tokamak science and technology base has been a great success story and will benefit the evolution of any magnetic concept. Improvements in the conventional tokamak as well as more significant departures, such as the Spherical Torus, may well lead to commercial success. ITER represents an opportunity to leverage off the mainstream of the world fusion effort in both fusion science and fusion technology. If ITER proceeds into construction, the US should attempt to be a significant participant."

The fifth topic discussed concerned the inertial confinement fusion pathway to commercial fusion.

The summary says, "The inertial fusion community has done outstanding science and technology work in the pursuit of its concepts and has provided a potential path to commercialization. However, additional work, not now pursued under weapons sponsorship, is required for commercial application. In view of the opportunity for leverage provided by the construction of the National Ignition Facility, the Office of Fusion Energy Sciences should actively develop the energy-related technologies for inertial fusion. All should strive to break down the barriers that separate the fusion community into inertial and magnetic camps."

The sixth topic discussed was "Promising non-tokamak magnetic pathways to commercial fusion."

The summary states, "There are a large number of concepts and variations on concepts and, despite FEAC and FESAC reviews of overall alternate concept policy, no agreed upon procedure is yet in place for setting priorities among them. Such a procedure should be established. Some alternates benefit from the extensive tokamak database more than others and hence progress on these concepts can be expedited. On the other hand, risk is reduced if some concepts explored are orthogonal to the toroidal magnetic concepts." The summary comments, "Although the DOE Office of Fusion Energy Sciences has announced plans to increase the level of effort on alternates, only modest increases have materialized thus far and the process for determining a new program balance is unclear. Though many concepts can apparently be explored at modest levels initially, the budgetary impact in future years of initiating a large number of concepts does not appear to have been thought through."

The final topic discussed was "Do fusion power plants really require low-activation materials?"

The summary concludes, "Materials are important to commercial success of many technologies, especially nuclear technologies. Fusion will require advanced materials that function in a unique environment. A materials development program is thus essential for fusion. Low-activation is a desirable attribute for fusion materials, but other attributes may be more important for both technical feasibility and economics. Therefore, a comprehensive program, which takes into account system tradeoffs is required."

Farewell to TFTR

1997 was the final year of operation of the major US fusion experiment, the Tokamak Fusion Test Reactor (TFTR), at the Princeton Plasma Physics Laboratory. TFTR was authorized in the FY 1976 budget and began operation in 1982 under the able leadership of Dale Meade and, later, Richard Hawryluk. It was shut down due to the Congressional budget cuts described above.

Among the many accomplishments of TFTR, it pioneered achieving the "high-temperature regime" of over 500 million degrees C; achieved record values of the Lawson triple product of density-temperature-confinement time; achieved in 1994 the first record values of fusion energy and power (over 6 million joules and almost 11 million watts); demonstrated "bootstrap current" which promises high-efficiency and steady-state tokamak power plants in the future; made the first measurements of the products of fusion reactions (helium nuclei) and demonstrated both their removal process and heating of the plasma by the fusion reaction products; and achieved routine operation with tritium fuel. All of the above were achieved in a declining budget atmosphere and in the face of a general lack of interest in energy programs in both the Congress and the Executive Branch after 1980. TFTR's sister facility, the Joint European Torus (JET), would later achieve 16 million watts of fusion power in 1997, and it continues to operate to the present day (2012).

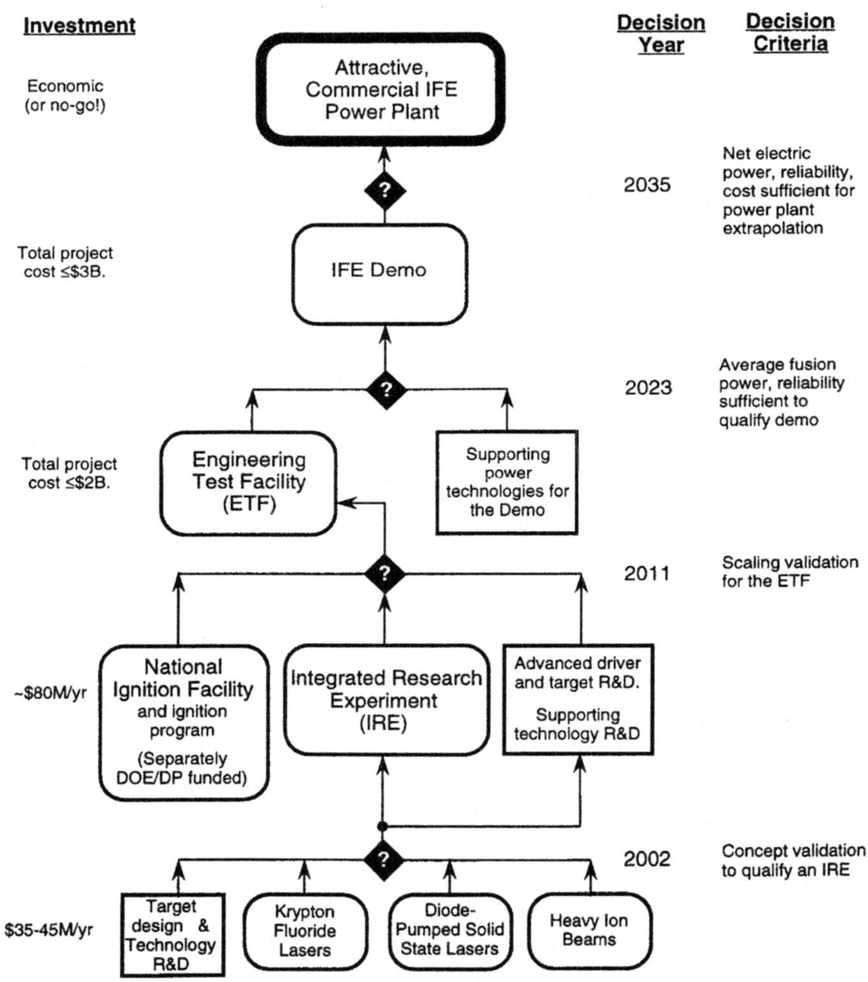

Fig. 8.2 Road map presented by inertial fusion advocates at workshop shows decision dates, facilities, and project costs leading to an inertial fusion power plant

Fusion Community Attempts to Regroup

During the week of April 27, 1998, about 150 persons, mostly from the magnetic fusion community, met for 1 week in Madison, Wisconsin, for a "Forum for Major Next-Step Experiments." However, members of the inertial fusion community at the forum, led by Mike Campbell (LLNL) and including Roger Bangerter (LBNL), Stephen Bodner (NRL), and Bob McCrory (U. Rochester), presented a comprehensive plan to develop a commercial fusion energy source (Fig. 8.2). They proposed a target date of 2023 for deciding to construct an inertial fusion (IFE) demonstration power plant.

The "road map" presented by the inertial fusion community stimulated the leaders of the magnetic fusion community to think in the same fashion, and soon they prepared a similar road map but notably without dates or costs (Fig. 8.3).

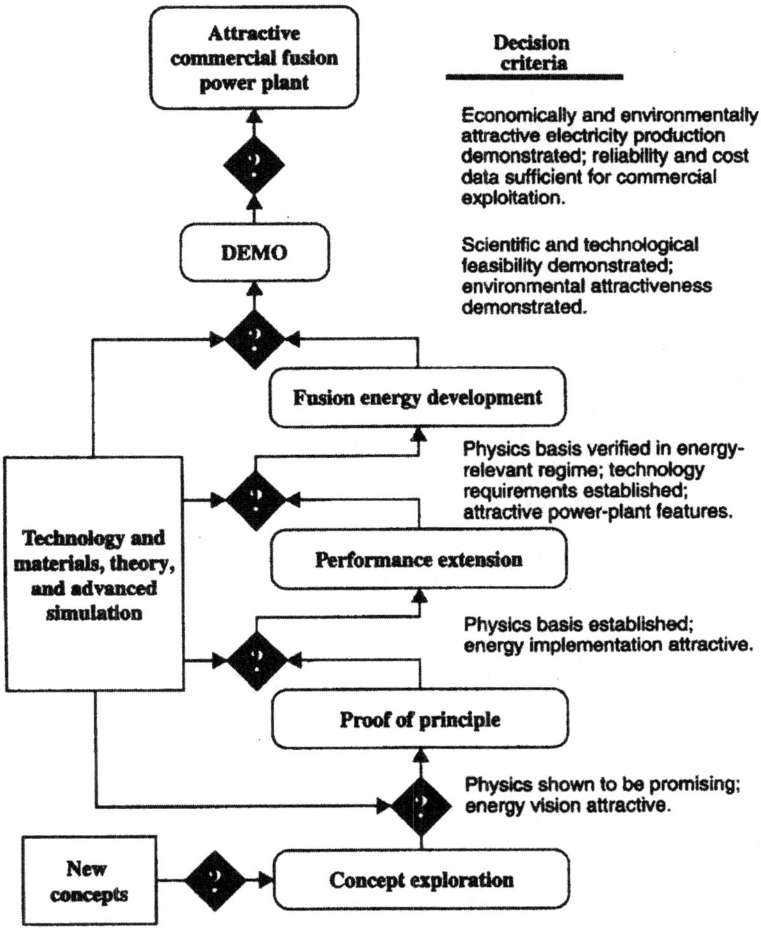

Fig. 8.3 Road map prepared by magnetic fusion advocates (notably without dates or costs) shows program phases leading to a commercial fusion power plant

A "next-step options" program was initiated within the US magnetic fusion program aimed at developing plans for initiating an experiment to study "burning plasmas."

Congress Orders USA Out of ITER Collaboration

During the summer and fall of 1998, the issue of continued US participation in the ITER project came to a head. The existing 6-year ITER engineering design activities (EDA) agreement expired in July 1998. The European Union, Japan, and Russia all signed a 3-year extension, but the USA was advised by Congress not to do so without Congressional approval. Language in the House-passed FY 1999 appropriations bill cautioned DOE that "The Congress has been very clear that no obligation exists for future participation in ITER beyond the Fiscal Year 1998 contribution for engineering design activities (EDA)." Congressman James Sensenbrenner (R-Wisconsin), chairman of the House Science Committee, took a personal interest in the ITER issue. He sent a letter to me dated August 7 in which he states, "I believe it is clear that ITER—as originally envisaged and specifically defined in the 1992 agreement—will not be built, and that there appears to be consensus among the ITER Parties that alternatives to the current design need to be explored. However, it is not clear that there is a consensus in either the US or the international fusion communities of what these alternatives should be or how a simple extension of the current ITER agreement—which was designed for a specific project—would accomplish this." Sensenbrenner's letter was in response to a July 25 letter I had sent to him stating, in part, "I have read your June 25 letter to then Secretary Peña urging him to delay signing the (ITER) extension agreement and I ask you to reconsider your view and to assist Under Secretary Ernest Moniz in his efforts to secure Congressional blessing on this agreement extension."

Congress did not back down, however. On September 22, 1998, Energy Secretary Bill Richardson, who had succeeded Secretary Peña, signed a 1-year extension of the ITER agreement, for the purpose of an orderly withdrawal of the USA from the ITER project. In the final passage of the FY 1999 appropriations bill, Congress calls for DOE "prior to committing to any future magnetic fusion program or facilities, conduct a broader review to determine which fusion technologies the US should pursue to achieve ignition or a fusion energy device."

FPA Meeting: Cost-Effective Steps to Fusion Power

Fusion Power Associates held its annual meeting and symposium January 25–27, 1999 [84]. Despite the signs, as described above, that the USA was pulling back from pursuit of a fusion energy goal via ITER, Dr. N. Anne Davies (Associate Director for Fusion Energy Sciences at DOE) told the audience, "While the U.S. fusion program has been restructured from fusion energy development to innovation-driven research focused on fusion's scientific foundations, we must preserve our long-term energy vision." She said, "The restructuring process has created opportunities to explore cost-effective pathways to this vision, with steps that are more affordable, deliver an improved fusion product, and provide the greatest return on investment of federal research funds." She noted, "There are trends in the marketplace that could make it difficult for fusion to compete, such as abundant supplies of fossil fuels, coupled with the possibility of sequestering carbon; declining net costs of producing electricity; and movement toward more distributed generation units in relatively small sizes. However, for the longer term, there are several factors that make us optimistic about the prospects for fusion." Among these factors, Davies noted, "One could imagine the emergence of large nuclear operating companies that have a long-term view and, because of a high comfort level with nuclear technologies, might embrace fusion and recycle former nuclear sites into fusion facilities." She also said that "environmental issues, such as climate change, will persist and perhaps grow more threatening with continued fossil fuel burning." And, she noted, "Global population growth, desire for standard of living improvements, and resource depletion issues strongly favor fusion in the long term." She said, "The viability of fusion in the marketplace will depend on its cost, reliability, and development path requirements relative to competing new energy sources."

In the late 1980s Y-K Martin Peng (ORNL) proposed the "Spherical Torus" concept, an approach to reducing the size and cost of tokamak fusion facilities. Basically, the idea is to shrink the size of the "hole in the donut" of the tokamak. Technically this is done by reducing the "aspect ratio" (ratio of the radius of the toroidal chamber to the radius of the confined plasma). Such a facility, the National Spherical Torus Experiment (NSTX) was nearing completion at PPPL in early 1999. At the FPA meeting, scientists from General Atomics laid out a 20-year fusion development program based on the Spherical Torus concept. Many other ideas, based on non-tokamak concepts, were also presented [84].

1998–1999 SEAB Review

In late 1998, Energy Secretary Bill Richardson (former United Nations Ambassador and later Governor of New Mexico) requested his Secretary of Energy Advisory Board (SEAB) to form a new fusion subcommittee to review the Department's fusion-related activities.

The SEAB Fusion Task Force, led by Richard Meserve (later to become chairman of the Nuclear Regulatory Commission), made its report on August 9, 1999 [85]. In transmitting the report to Secretary Richardson, SEAB chairman Andrew Athy said that "the fusion program must be led by strong management, capable of directing the program towards its goals at a reasonable pace" and with a sufficient budget, "on the order of $300 million per year." Strong management was the antithesis of what the magnetic fusion scientists established following its "restructuring" and subsequent transition to a "science" focus just 2 years earlier.

The SEAB report said:

> "It is the Task Force's view that the threshold scientific question—namely, whether a fusion reaction producing sufficient net energy gain to be attractive as a commercial power source can

> be sustained and controlled—can and will be solved. The time when this achievement will be accomplished is dependent, among other factors, on the creativity of scientists and engineers, skill in management, the adequacy of funding, and effectiveness of international cooperation."
>
> "Nonetheless, there remain significant barriers to the realization of fusion as a significant contributor to the world's energy supply. Progress requires advancing fundamental scientific knowledge (from controlling turbulence, to optimizing the magnetic-field configuration, to enhancing the fusion power gain), resolving very difficult materials issues (e.g., developing a vessel that can withstand high temperatures and intense neutron flux), finding answers to difficult engineering challenges (e.g., constructing a reliable and repairable system), and proving economic feasibility (solving these problems in a manner that does not make fusion prohibitively expensive). Many years of persistent effort will be required to overcome these challenges. In spite of the extended effort and expense that will be required, the fusion program deserves continued support because of its unique energy potential. Constraints on supply and limits on the atmospheric loading of combustion products will eventually require that we diminish our reliance on fossil fuels. Because of this reality, the Department is wisely advancing a portfolio of energy technologies to meet future energy needs. Indeed, in light of the promise of fusion and the risks arising from increasing worldwide energy demand and from eventually declining fossil energy supply, we simply cannot afford to fail to pursue fusion energy aggressively."

With respect to the magnetic fusion energy (MFE) program, the Task Force said that they endorse the "revised focus of the program" away from a "nearly exclusive focus on the achievement of fusion energy in tokamaks to a broader program that would also explore scientific foundations and other confinement approaches." They said "OFES (DOE Office of Fusion Energy Sciences) has begun to expand the fusion portfolio and it should be encouraged to continue this effort." They said, "It is our view that the Department must participate in international activities that enhance our fusion effort. Communication with the Congress on these points is essential."

With respect to inertial fusion energy (IFE), the Task Force said, "As is the case for MFE, progress in inertial fusion has been remarkable. The scientific basis of inertial fusion has progressed to the point where the driver and pellet requirements to achieve ignition are known to high confidence and are within reach." The Task Force noted that "Some considerations favor heavy ion beams as the driver technology for IFE." But, they said, "Given the immature state of the technology, it is not appropriate at this time to select only one driver technology for continued exploration." They also said that reactor studies "should continue to be used as guides in establishing the direction and balance of research efforts, as well as to establish goals that constitute thresholds for further investment."

In the report, the Task Force says, "Efforts to define a sensible path leading to a substantial energy contribution from fusion should be given continuing emphasis." They say, "It is crucial that this planning encompass the identification and timely resolution of the important engineering and economic problems that must be overcome if fusion energy is to be a practical energy source." The report says, "Management should seek to restore credibility by articulating clear and sensible milestones and goals and to deliver on them." It states, "Given constrained budgets, the wide variety of options, and the linkages of one issue to another, increasingly sophisticated management of the program will be required." The Task Force report calls for "the application of new management tools and techniques." It says, "Given the complex nature of the fusion effort, an integrated program planning process is an absolute necessity." It says, "Proper management of the fusion program requires a comprehensive planning system that: provides visibility of program activities; provides the means to manage by performance; encourages fundamental, innovative scientific research; drives resource planning; provides linkage of accomplishments to goals; establishes accountability; encourages the development of trained personnel; describes activity interrelationships, and aids in integration among the base programs in OFES and DP and the fusion energy goal of practical fusion energy."

DOE, however, had no interest in establishing a strong management style for fusion, preferring to treat it as they did their basic (open-ended) scientific research programs.

1999 Fusion Snowmass Meeting

In July 1999, more than 300 physicists from across the United States and 11 other countries met for 2 weeks in Snowmass Village, Colorado, to discuss the present state of the US Fusion Energy Sciences research program and its future direction. The long, formal title of this meeting was "1999 Fusion Summer Study: Opportunities and Directions in Fusion Energy Science for the Next Decade." Importantly, the magnetic confinement effort and the inertial confinement effort were both broadly represented. Making specific decisions about program direction was not in the charter of the meeting; however, the work accomplished had a significant effect on the direction of the US fusion program.

FESAC Comments on Fusion Program Priorities and Balance

At its September 8–9, 1999 meeting, the DOE's Fusion Energy Sciences Advisory Committee (FESAC) sent recommendations to DOE Office of Science Director Martha Krebs on fusion program "priorities and balance." A key recommendation was that, even at the present $222 million level of funding, some $5 million should be shifted from magnetic fusion energy (MFE) to inertial fusion energy (IFE). The FESAC said, "The achievement of a more integrated national program in MFE and IFE (should be) a major programmatic and policy goal in the years ahead." They said the MFE program "is currently reasonably well balanced among its programmatic sub-elements," saying that the program "properly emphasizes steady-state, externally-controlled configurations, such as the advanced tokamak and spherical torus." However, they said, "To maintain a proper balance, care must be taken to also maintain an emphasis on pulsed and/or self-organized concepts." They referenced a report, "Opportunities in the Fusion Energy Sciences Program" prepared in June 1999 under the direction of FESAC chair John Sheffield.

Martha Krebs stepped down from her post in early December 1999.

OMB View

Despite the obvious interest in fusion energy applications by both the SEAB, PCAST, and most members of the US fusion community, fusion continued to be viewed as a "science program" at the Office of Management and Budget. Speaking at Fusion Power Associates annual meeting on October 19, 1999 [86], OMB fusion budget examiner Dr. Michael Holland said, "From OMB's view, I'd like to emphasize that we see fusion as a science program and not an energy technology program. And that means that we judge you according to the criteria that we judge the other programs in the science portfolio: high energy physics, nuclear physics, basic energy sciences. Scientific excellence is the critical performance measure that we look for. Part of the reason why we look at fusion sciences as a science program and not an energy technology program is due to some of the recent actions that Congress took, particularly moving fusion out of the energy supply budget account and into the science account." In response to questions, Dr. Holland made the following additional comments: "My personal feeling is that the technology aspects of the fusion sciences program ought to be considered in the same way that the technology aspects of high energy physics are considered. We invest a lot in accelerator R&D, but we do that to advance science in high-energy physics. And accelerator R&D is not an end to itself. So if the technology aspects of the fusion sciences program are connected to the science that you're trying to advance, then I think that's a wise investment. I guess that's the only way I would imagine doing that part of the budget."

As the decade (and the century) came to an end, the MFE fusion budget remained at around $230 million, well below the $345 million level it was at 10 years earlier. The inertial confinement fusion budget within DOE's weapons category, on the other hand, was at $508 million in 1999, compared to its 1989 level of only $164 million. This was due to the decision to construct the laser-based National Ignition Facility (NIF). However, within the total provided for inertial confinement fusion, the Congress began in 1998 providing $25 million per year for a High Average Power Laser (HAPL) program aimed at the development of civilian energy applications. For the next 10 years, the DOE weapons program steadfastly refused to ask for funds for this "civilian" effort, stating that it was outside the scope of their weapons mission and the Congress just as steadfastly continued to provide the money and the charge to pursue the energy applications of inertial fusion. The HAPL program was led by John Sethian at the US Naval Research Laboratory but was national in scope. It employed an integrated systems approach to developing all the subsystems necessary for a laser-based, inertial fusion power plant.

No Fusion Demonstration Power Plant by 2000

As the twentieth century drew to a close, it was clear that the world would not have an operating fusion power plant, as envisaged in the 1976 plan [24] by the year 2000. The reason was threefold: failure of the US government to provide

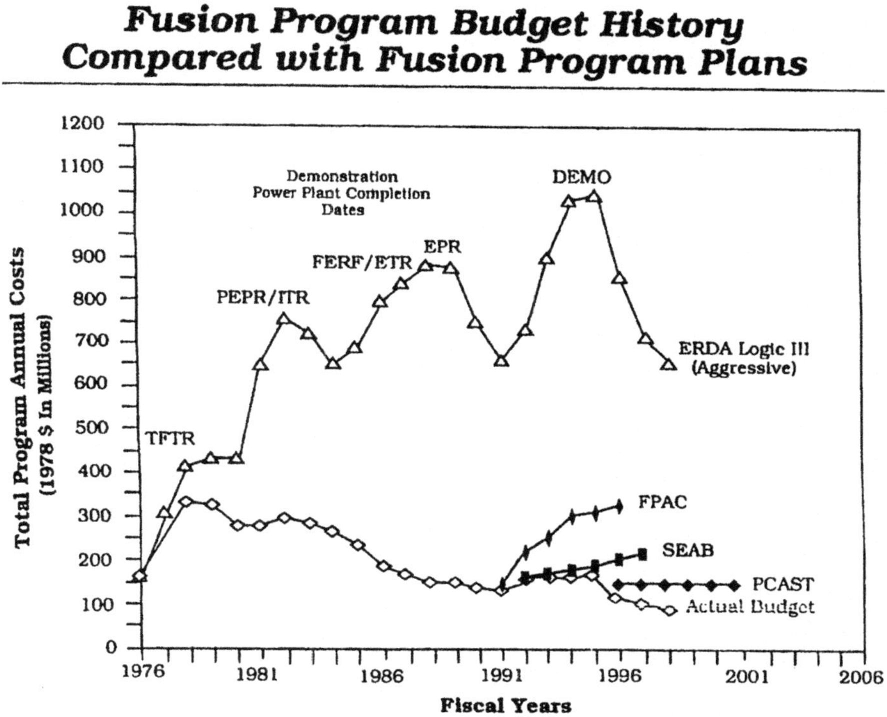

Fig. 8.4 Funds required to achieve a demonstration power plant by the year 2000, as laid out in Logic III of the 1976 plan (demonstration reactor by 1998), compared to the actual funds provided through 1998. Also shown on the figure are recommendations of the DOE's Fusion Policy Advisory Committee (FPAC), the Secretary of Energy Advisory Board (SEAB), and a level-funding proposal by PCAST. The facilities labeled as needed in the 1976 plan after TFTR were a prototype experimental power reactor or ignition test reactor (PEPR/ITR), a fusion engineering research facility or engineering test reactor (FEFF/ETR), an experimental power reactor (EPR), and a demonstration power plant (DEMO)

the necessary funds (Fig. 8.4); failure of the US government to commit to construction of necessary facilities, such as the fusion engineering test reactor; and failure of the US Department of Energy to manage the program so as to achieve its avowed practical purpose.

The New Millennium: Science vs. Energy: 2000–2008 9

The difference between a goal and a wish is that a goal has a definite time frame attached to it, within which we intend to do certain things to help accomplish that goal.

Thomas W. McKnight
Author

In the spring of 2000, MSNBC carried out an online poll about fusion. As of March 27, 8,000 persons had responded. Of these, 65% said they believed fusion "will make a difference in my lifetime;" 25% believed fusion "will become an important power source, but not in my lifetime;" and 6% said fusion "will not make a difference" due to either "economic or scientific" factors.

Although DOE Office of Fusion Energy Sciences head N. Anne Davies had indicated at the September 1999 FPA meeting her intention to keep the fusion program on a path towards an energy goal, her efforts would meet stiff resistance, as we shall see.

Fusion Power Associates had just celebrated its 20-year anniversary. The original motivation of FPA was to mobilize the private sector for fusion engineering and commercial deployment. As indicated in Table 5.3, FPA initially had the support of many industries and electric utilities. As it became clear, first by deregulation of the electric utility industry in the 1980s and then by the elimination of industrial opportunities to participate in fusion R&D due to lack of new fusion facility construction and US pullout from ITER, many private sector companies gradually withdrew from direct participation in FPA. The major national laboratories and universities involved in fusion research began to join FPA, however, and soon became our major supporters and participants, as indicated in Table 9.1. DOE did provide some recognition to FPA, however. In July 2000, Energy Secretary Bill Richardson presented me with my second DOE Distinguished Associate Award. The award states, "United States Department of Energy Distinguished Associate Award presented to Dr. Stephen O. Dean, for your many contributions to the fusion program over the past 20 years in your role as the co-founder and President of Fusion Power Associates. Your tireless work in bringing diverse components of the fusion community together for topical meetings, and your efforts to increase public awareness and understanding

S.O. Dean, *Search for the Ultimate Energy Source: A History of the U.S. Fusion Energy Program*, Green Energy and Technology, DOI 10.1007/978-1-4614-6037-4_9,

Table 9.1 Fusion Power Associates Institutional Members, 2012

The Boeing Company	University of California, Los Angeles
Columbia University	University of California, San Diego
General Atomics	University of Rochester
Lawrence Berkeley National Laboratory	University of Texas, Austin
Lawrence Livermore National Laboratory	University of Wisconsin
Los Alamos National Laboratory	Calabazas Creek Research, Inc.
Massachusetts Institute of Technology	General Fusion
Oak Ridge National Laboratory	Logos Technologies
Princeton Plasma Physics Laboratory	Tech-X Corporation
Sandia National Laboratories	Woodruff Scientific, Inc.
Savannah River National Laboratory	
Schafer Corporation	

of the potential benefits of fusion, have helped to make fusion an important element of the Nation's energy sciences portfolio."

National Energy Policy Development Group

In November 2000, George W. Bush, son of previous President George H. W. Bush, won the US presidential election, succeeding President Bill Clinton. One of his first acts, which gave a boost to the hopes of the US fusion community, was to establish a National Energy Policy Development Group under the direction of Vice President Dick Cheney. I prepared a statement on Fusion and Energy Policy, ultimately signed by 190 persons from both within and outside the fusion community [87] and, on February 5, 2001, sent this to the Energy Policy Group. The statement urged the group, "In addition to acting to ameliorate the immediate problems, we urge you to address the nation's long-term energy needs through creation and funding of a focused R&D effort to expand our future commercial energy options." The statement said, "We advocate an expanded, sustained energy R&D effort to provide the United States and the world with the energy it will need for the twenty-first century. The focus of this effort should be to provide new economic and environmentally acceptable energy technologies for improving energy end-use efficiency, cleaner burning of fossil fuels, improvements in nuclear fission technologies, and less costly and more efficient renewable energy options. For the long-term, we urge an accelerated effort to develop fusion energy." In a March 26 response letter, the new Energy Secretary, Spencer Abraham, said, "As I noted at my confirmation hearing in January, I am deeply committed to developing an energy policy that includes increasing domestic production of energy in an environmentally responsible manner, increasing our use of renewable energy, decreasing our reliance on imported oil, and developing new technologies that will reduce energy-related pollution." He said, "The (DOE) Office of Fusion Energy Sciences' work to provide the knowledge needed for an economically and environmentally attractive source of energy and to advance our understanding of plasma science and fusion science is an important part of our efforts in science and technology." Treasury Secretary Paul O'Neill, another member of the Energy Policy Group, also responded saying, "I agree that energy supply is a long-term issue and that R&D is key to the long-term health of the industry, and I am familiar with the promise of fusion power. We will keep this in mind in our coming efforts to craft an effective forward-looking energy policy."

On May 17, 2001, President Bush issued the report of his National Energy Policy Development (NEPD) Group. Although the report focused on near-term and mid-term energy sources, the report also comments on fusion, saying, "The NEPD Group recommends that the President direct the Secretary of Energy to develop next generation technology—including hydrogen and fusion."

The report states:

> "Fusion—the energy source of the sun—has the long-range potential to serve as an abundant and clean source of energy. The basic fuels, deuterium (a heavy form of hydrogen) and lithium, are abundantly available to all nations for thousands of years. There are no emissions from fusion. The radioactive wastes from fusion are short-lived, only requiring burial and oversight for about 100 years. In addition, there is no risk of a meltdown accident because only a small amount of fuel is present in the system at any time. Finally, there is little risk of nuclear proliferation because special nuclear materials, such as uranium and plutonium, are not required for fusion energy. Fusion systems could power an energy supply chain based on hydrogen and fuel cells, as well as provide electricity directly."
>
> "Although still in its early stages of development, fusion research has made some advances. In the early 1970s, fusion research achieved the milestone of producing one-tenth watt of fusion power for one-hundredth of a second. Today, the energy produced from fusion is ten billion times greater, and has been demonstrated in the laboratory at powers over 10 million watts in the range of a second."
>
> "Internationally, an effort is underway in Europe, Japan and Russia to develop plans for constructing a large-scale fusion science and engineering test facility. This test facility may someday be capable of steady operation with fusion power in the range of hundreds of megawatts."
>
> "Both hydrogen and fusion must make significant progress before they can become viable sources of energy. However, the technological advances experienced over the past decade and the advances yet to come will hopefully transform the energy sources of the distant future."

National Academies Report on the Quality of Fusion Science

In the spring of 2001, the National Academies of Science and Engineering issued a report on the quality of the science in the US fusion program. DOE had requested the report in April 1998. The final report [88] titled "An Assessment of the Department of Energy's Office of Fusion Energy Sciences Program" states, "The Committee concludes, therefore, that the quality of the science funded by the United States fusion research program in pursuit of a practical source of power from fusion (the fusion energy goal) is easily on a par with the quality in other leading areas of contemporary physical science." Dr. Charles Kennel, Director of the Scripps Institution of Oceanography and a former Deputy Director of NASA, chaired the committee.

Although the title of the report might lead one to believe that the entire fusion program was assessed, in fact its scope was limited to fusion science. The report acknowledges in its Preface: "The report focuses on the science of magnetically confined plasma and the programmatic strategy for long-term progress in this area, but it does not directly address inertially confined plasma . . . . Also, this assessment does not directly address issues surrounding specific technology development and engineering research sponsored by the program. . . because the committee chose to focus on elements of the program related to basic physics research." Nevertheless, the committee threw in the seemingly gratuitous remark: "A strong case can also be made that a program organized around critical science goals will also maximize progress toward a practical fusion power source." Nowhere in the report do they attempt to make that case, however. On the contrary, the report acknowledges that "Consonant with its charge, the committee has not taken up the many critical path issues associated with basic technology development for fusion, nor has it looked at the engineering of fusion energy devices and power plants, yet it is the combined progress made in science and engineering that will determine the pace of advancement toward the energy goal."

Burning Plasma Physics

As the decade began, the study of "burning plasma physics" became more and more stated as the focus of the US fusion effort, rather than on a balanced science and technology effort aimed at a fusion power plant.

Mildred Dresselhaus, a solid-state physicist from MIT, replaced Martha Krebs as Director of the DOE Office of Science, and Richard Hazeltine, a fusion plasma theoretical physicist from the University of Texas, replaced John Sheffield (ORNL) as chair of the FESAC.

The University Fusion Association (UFA) held workshops on Burning Plasma Science at the University of Texas (December 11–13, 2000) and at General Atomics (May 1–2, 2001). Prime movers for these workshops included Gerald Navratil (Columbia University), Ronald Parker (MIT), Mohamed Abdou (UCLA), and Farrokh Najmabadi (UCSD). The workshops addressed both the science issues associated with burning plasma physics and the technology requirements for building a facility that could be used for doing such experiments by the end of the decade.

Following the workshops, the UFA issued the following statement:

> "A burning plasma (BP) experiment would greatly strengthen the US fusion energy sciences program. The TFTR and JET experiments have produced reactor-like plasmas and attained near breakeven conditions ($Q \sim 1$). The alpha particle beta and energetic particle heating effects in these experiments were reactor-like, allowing the first exploration of BP physics. These $Q \sim 1$ results using a tokamak magnetic configuration give high confidence in the feasibility of a $Q > 5$ experiment. Production of a strongly self-heated fusion plasma will allow the study of a number of new phenomena. The non-linear coupling between

fusion alpha production, alpha-heating-sustained pressure transport, and boundary plasma behavior, as well as fusion ignition transient phenomena will be studied and controlled. The additional studies of Alfven wave dynamics, the effect of energetic particles on collisionless reconnection and proton and alpha particle heating will also impact space and astrophysical plasma physics."

"While fusion research is ready for a BP experiment, more knowledge of plasma physics is required for building a cost-effective reactor. Innovations to improve the economics of the tokamak or a more cost-effective configuration are needed for a practical reactor. A BP experiment will open up new scientific frontiers of study and take us a critical step closer to realizing the goal of fusion power. Further, clearly demonstrating that a BP can be achieved in a tokamak configuration will be directly applicable to a large number of related magnetic configurations. In addition, achievement will stimulate the creative engineering and technical development needed to make fusion practical. Finally, operating in high-Q regimes allows new discoveries, leading to significant advances towards practical fusion energy. The UFA supports the exploration of potential BP experiments and advocates that this important next step be pursued by the US fusion energy sciences program."

"The main focus of the US fusion energy sciences program is to develop the science and technical base needed for practical fusion energy by exploration across a broad spectrum of magnetic configurations. Each innovative confinement concept being investigated offers advantages that would improve the economics and/or reliability of a fusion power system. Also pursued in the present program are basic plasma science, plasma theory, computational plasma physics, system studies, and technology research that are essential to develop new understanding that leads to progress toward practical fusion and towards other applications of plasmas. This base program is needed to advance essential science and technology, to develop a more cost-effective concept, and to capitalize on advances made with a burning plasma experiment. Thus, a BP experiment must be funded with a significant augmentation of the fusion budget. The relatively flat funding in the US fusion energy sciences program for the past several years, following a major budget cut in1996, has left the base program badly underfunded. Therefore, the UFA supports a balanced program for a faster realization of fusion power, requiring an increased base program as well as a BP experiment."

In October 2000, Dresselhaus had asked the FESAC "to address the scientific issues of burning plasma physics." She said, "For many years, the U.S. magnetic fusion community has recognized that burning plasma physics is the next frontier of fusion research," and "in the last two decades, the program has made several attempts, both international and domestic, to move forward on the design and construction of a tokamak experimental device in which the science of burning plasmas could be explored." "For various reasons, all these attempts failed," she noted. She asked for a report by July 21, 2001.

Prof. Jeffrey Freidberg (MIT) chaired the FESAC "Burning Plasma Panel." Its report was endorsed by the FESAC and transmitted to DOE in late 2001. The FESAC said, "FESAC fully endorses the recommendations of the Burning Plasma Panel. In particular, we agree with the Panel recommendation that a burning plasma experiment would bring enormous scientific and technical rewards. We also agree that present scientific understanding and technical expertise allow confidence that such an experiment, however challenging, would succeed." The FESAC recommended that the fusion community hold a "Snowmass" workshop in the summer of 2002 "for critical scientific and technical examination of proposed burning plasma experimental designs."

The renewed interest in the USA in constructing a burning plasma physics experiment was, in part, motivated by uncertainties on whether ITER would actually go into construction and the fact that the USA had withdrawn from the ITER partnership. The mood in Congress and the Administration regarding ITER was changing, however.

Major Advance: DIII-D

On July 2, 2001, scientists working on the DIII-D tokamak at General Atomics announced a major scientific advance, saying they had "nearly doubled the usual limits on pressure in a fusion energy device by spinning the hot fusion fuel very rapidly." They noted, "High pressure in the fusion fuel is critical because the power released from fusion reactions increases very rapidly with increasing pressure." They said, "These results are an important step towards controlled fusion power production that is feasible, economical and attractive." Scientists from Columbia University and Princeton Plasma Physics Laboratory collaborated with the GA scientists on the experiments.

ITER Rising?

On November 1, 2001, Sherwood Boehlert (R-NY), chair of the House Science Committee, and subcommittee chair Ralph Hall (R-TX) wrote to Energy Secretary Spencer Abraham "to urge you to begin sending a representative to the international discussions regarding the International Thermonuclear Experimental Reactor (ITER) which, as you know, is a major fusion research initiative." The congressmen stated, "Obviously time is of the essence with the ITER initiative, and the U.S. should begin to assess the project's feasibility, evaluate what role the U.S. might play in it, and participate in discussions to refine the project and select a site." They said, "If we do not begin to examine ITER soon, we may lose the chance to join as a partner."

About the same time, President George W. Bush named Raymond L. Orbach to be Director of DOE Office of Science. At the time, Orbach, a condensed matter physicist, was Chancellor of the University of California at Riverside. He would soon become a champion for ITER. During 2001, as the President and his Science Advisor, John Marburger, went to meetings with their international counterparts in Europe, Japan, and Russia, they were continually asked why the USA was not participating in ITER. Naturally, this resulted in a reexamination of the US stance vis-à-vis ITER participation.

On January 3, 2002, Energy Secretary Abraham responded to the November 1, 2001, letter he had received from representatives Boehlert and Hall. He said, "I have agreed to explore the current ITER option before us to determine if it is appropriate for the Department—and the Nation—in the light of the President's National Energy Policy." He noted, "Representatives of other governments have asked that the Department review its current policy towards ITER." In a related development, the

British science journal, Nature, editorialized, "Any real exploration of long-term, sustainable energy must include a thorough technical evaluation of nuclear fusion." They said, "Bush would do well to buy back into ITER, and to help build it as a model of scientific collaboration." In the same issue of Nature, science writer Geoff Brumfiel quotes Bush's Science Advisor John Marburger as saying, "I definitely think that our participation (in ITER) should be reconsidered."

Speaking at a conference of G8 Energy Ministers, May 2, 2002, in Detroit, Energy Secretary Spencer Abraham said in part, "Advanced technology, and the contributions of science, play a critical role in our future energy plans. Along with promising innovations such as hydrogen fuel cells, the President is anxious to accelerate fusion power as a realistic source of energy. We are now engaged in serious consultation here in the United States and around the world on how best to pursue a fusion program. President Bush is particularly interested in the potential of the international effort known as ITER and has asked us to seriously consider American participation. This major international effort will answer a critical scientific question: Can a fusion reaction—the kind of reaction that powers the sun—be harnessed here on earth for the benefit of all mankind?"

Fifty Years of US Fusion Research

The USA formally began its fusion program, Project Sherwood, in 1952. In early 2002, the editors of the American Nuclear Society magazine *Nuclear News* asked me to write an article summarizing the 50-year history of fusion research in the USA. My article was published in the July 2002 issue [89]. In it I commented, "Fusion has been certified as ready for engineering development for more than 20 years, but a weak-willed government has been unwilling to manage and fund the program to accomplish its avowed practical purpose."

The High Average Power Laser (HAPL) Program

Beginning in late 1998, Congress continually urged the DOE to fund research aimed at civilian fusion power via inertial confinement. Inertial confinement fusion was funded and managed within the DOE's weapons program, which steadfastly asserted that such a civilian application was not within their "mission." Civilian applications of inertial confinement fusion require that small capsules of fusion fuel be ignited repetitively (perhaps ten times per second) and not once or a few times per day, which is sufficient for weapons data. Although several methods (called "drivers") are promising for this application (heavy-ion accelerators, Z-pinch-produced X-rays, and lasers), Congress chose the laser driver and provided approximately $25 million per year beginning in FY 1999 for the next decade to the DOE weapons program. Each subsequent year DOE failed to request funds to continue the effort, and each year the

Congress added funds for it. The program was called HAPL. It was a national effort, brilliantly coordinated by John Sethian at the US Naval Research Laboratory.

Three HAPL workshops were held in 2001; another in early 2002. At the April 4–5, 2002, workshop, Sethian described the program as a "coordinated, focused, multi-lab effort to develop the science and technology for laser fusion energy, based on lasers, direct-drive targets and solid-wall chambers." He said the program was one in which a "systems approach" was being used so that key components would be developed together. This, in fact, contrasted sharply with the management approach of the magnetic confinement fusion program in the DOE Office of Science, which beginning in the mid- to late-1990s, had adopted a serial approach of focusing first on the science.

The 2002 Fusion Summer Study

Over 280 fusion researchers (mostly from the magnetic community), including over 30 non-US participants, met in Snowmass, Colorado, for 2 weeks in mid-July 2002. The focus of the workshop was on identifying "next step" facility options, especially those aimed at the study of "burning plasmas." Three magnetic fusion facilities had been proposed: a high-field, copper magnet tokamak called Ignitor, an advanced high-field copper magnet tokamak called FIRE, and the international ITER engineering test reactor. Progress and plans in the area of inertial fusion for energy were also discussed. The workshop group favored either FIRE or ITER, with FIRE viewed as a primarily US facility with lower cost and less reactor features compared to ITER, which was viewed as an international facility.

For inertial fusion, the Summer Study noted that the National Ignition Facility (NIF), already under construction at the Lawrence Livermore National Laboratory, "is expected to produce a burning inertial fusion plasma."

Following the Summer Study, the DOE FESAC convened a special panel, of which I was a member, to consider further the magnetic burning plasma facility options. The panel met August 6–8 in Austin, Texas. A formal meeting of the full FESAC followed that meeting, September 11–12 in Gaithersburg, Maryland.

The Austin panel report was endorsed by the full FESAC. The Austin panel based its recommendation on an estimated the total cost of ITER construction of $5 billion, with the US contribution to be $500 million. The panel favored ITER over FIRE, provided the US contribution to ITER came in the form of growth in the present US fusion budget. With that assumption, the panel report stated, "we should now seek to join the ITER negotiations with the aim of becoming a partner in the undertaking." The panel and the FESAC said, "The desired role is that the US participates as a partner in the full range of activities, including full participation in the governance of the project and the program." FESAC chair Richard Hazeltine transmitted the FESAC recommendation to Raymond Orbach, the new Director of the DOE Office of Science.

As the Administration considered whether the USA should rejoin ITER, Orbach and John Marburger, Science Advisor to President George Bush, asked the FESAC

Table 9.2 FESAC 35-year Plan Panel

Robert J. Goldston, Chair	B. Grant Logan
Mohamed Abdou	Kathryn McCarthy
Charles Baker	Farrokh Najmabadi
E. Michael Campbell	Craig Olson
Vincent Chan	Stewart Prager
Stephen O. Dean	Ned Sauthoff
Amanda Hubbard	John Sethian
Robert Iotti	John Sheffield
Thomas Jarboe	Steven Zinkle
John Lindl	

to place ITER in the context of an energy development strategy. At the September 11–12, 2002, FESAC meeting, Orbach asked the committee "to develop a plan with the end goal of the start of operation of a demonstration power plant in approximately 35 years." The plan, he said, "should recognize the capabilities of all fusion facilities around the world, and include both magnetic fusion energy (MFE) and inertial fusion energy (IFE), as both MFE and IFE provide major opportunities for moving forward with fusion energy." Orbach said he wanted a preliminary plan within 2 months because he wanted "to give the President by mid-December the full scientific picture of how to get from here to there (fusion power)" and "a final detailed plan by March 2003." He said, "These are historic times for the fusion program."

FESAC set up a panel to prepare the fusion energy plan. PPPL Director Rob Goldston chaired the panel, of which I was one of the members. The full membership is shown in Table 9.2. The panel held meetings on October 3–4 at Princeton, October 28–31 in Livermore, and November 15–17 in Orlando, Florida. The panel provided its interim report to FESAC at its November 25–26, 2002, meeting in Gaithersburg, Maryland.

The fusion community was enthusiastic about Orbach's comments and his request for a plan to achieve the energy goal. They hoped this marked a transition back to an energy goal with a timetable to a demonstration power plant. Alas, they were (once again) to be disappointed.

Another Academies Fusion Review Panel

A new review of the fusion program by the National Academies began with a September 17–18, 2002, meeting in Washington, DC. The topic was a scientific-oriented one. The name of the new committee was "Burning Plasma Assessment Committee (BPAC)." It was cochaired by John Ahearne (Sigma Xi Center) and Raymond Fonck (a plasma physicist from the University of Wisconsin, who would later become director of the US fusion program at DOE).

On November 18, 2002, the President's Science Advisor, John Marburger, addressed the BPAC "regarding the Administration's view of issues associated with the peaceful exploration of nuclear fusion." He said, "Let me say at the outset that this Administration is supportive of the concept of electrical power generation from nuclear fusion." He commented, "The closer we are to a transition from a fusion science program to a fusion device engineering program, the easier it will be to create favorable economic conditions to accelerate the practical implementation of fusion power." He said, "The promise of fusion is too great to ignore—but we also understand that this has been a true statement for 50 years." Marburger said, "I believe the fusion community has made a compelling case that a burning plasma experiment is the essential next scientific step for fusion research. I am convinced there is no foreseeable path to practical fusion without a burning plasma experiment." He added, "(an) equally important part is the search for a commercially optimal containment technology." He said, "Other issues, like the development of materials that can withstand 14 MeV neutrons or the design of blanket technologies are only important once the first problem—the creation of a burning plasma—has been solved."

As part of the process of deciding whether to rejoin ITER, Orbach commissioned an internal DOE review of the ITER cost estimate. The DOE review committee "concluded that the ITER team has prepared a complete cost estimate that is based on sound management and engineering principles, and is credible as a basis for establishing relative contributions by the Parties to the construction of ITER." The committee said the ITER cost estimate, which they estimated at $5 billion (in constant 2002 dollars), "is supported by the design and R&D results that are unusually mature for a science project facing the decision to fund construction."

USA Rejoins ITER

In January 2003, President George W. Bush issued a statement saying, "I am pleased to announce that the United States will join ITER, an ambitious international research project to harness the promise of fusion energy. The results of ITER will advance the effort to produce clean, safe, renewable and commercially available fusion energy by the middle of this century. Commercialization of fusion has the potential to dramatically improve America's energy security while significantly reducing air pollution and emissions of greenhouse gases." A DOE press release stated, "The U.S. share of the (estimated $5 billion) construction cost is expected to be about 10% of the total." In a speech to fusion researchers at PPPL on January 30, Energy Secretary Abraham said, "But let me be clear, our decision to join ITER in no way means a lesser role for the fusion programs we undertake here at home. It is imperative that we maintain and enhance our strong domestic research program—at Princeton, at the universities and at our other labs. Critical science needs to be done in the US, in parallel with ITER, to strengthen our competitive position in fusion technology."

The 35-Year Plan

In March 2003, the FESAC panel chaired by Rob Goldston delivered the "35-Year Plan" previously requested by DOE Office of Science Director Ray Orbach. The plan had also been requested by the President's Science Advisor, John Marburger, who had been told by his European counterparts of the existence of such a plan within the European fusion effort. The FESAC plan envisaged a broad portfolio of both magnetic and inertial fusion energy approaches and associated technologies over the next 15 years at a total cost of approximately $10 billion. At that time the concept for the first generation of fusion power plants would be selected for focused development over the next approximately 20 years [90].

The plan called for initiating the development effort in FY 2004 with a budget of $332 million, although the recently submitted Presidential FY 2004 budget requested only $257 million. For its implementation, the plan required the fusion budget to continue to grow to approximately $570 million in 2008 and to peak at approximately $900 million around 2013. The plan states, "To achieve the goals of this plan, the program must be directed by strong management. Given constrained budgets, the wide variety of options and the linkages of one issue with another, increasingly sophisticated management of the program will be required."

When the plan was requested, the fusion community thought the Administration was seriously interested in fusion energy development. We were told that such a plan was necessary so that the President would understand the context for the decision he was being asked to make to rejoin ITER. It turned out, however, that the ITER decision was made prior to completion of the plan. Hence, when the plan was delivered, it was deemed unnecessary and was subsequently ignored. Worse, the DOE OFES FY 2004 budget document, though it contained $12 million to restart the US ITER effort, eliminated essentially all the energy-oriented fusion technology programs.

At its meeting on March 5–6, 2003, the FESAC expressed dismay at the fusion technology cuts proposed in the FY 2004 budget submission. In a letter dated March 5, FESAC chairman Richard Hazeltine told DOE Office of Science Director Ray Orbach that "devastating cuts to certain program elements are alarming; this note expresses our most serious concerns." Hazeltine said, "The fusion energy sciences budget for FY 2004, as described in the President's request, stunned FESAC members." He said, "FESAC is puzzled by the elimination in FY 2004 budget of funding for fusion technology. This loss will seriously compromise US participation in ITER as well as other burning plasma research activities." He added, "The study of future energy systems is a central component of fusion research. Its evolving conceptualization of an eventual fusion power plant has helped us visualize our target, while allowing us to identify key scientific challenges. As the energy goal becomes closer and more central to fusion research, such systems studies provide even more important insights. Yet the FY 2004 budget significantly reduces funding for this type of research." Hazeltine further noted, "FESAC recommendations regarding the burning plasma initiative have emphasized the importance of maintaining scientific and technological breadth in the program. The Secretary of Energy renewed this emphasis in his recent announcement concerning US participation in ITER. Yet funding for FIRE, a domestic burning plasma experiment that could provide an alternative to ITER, has been eliminated. Similarly, inertial fusion energy (IFE) is an important element of a balanced US fusion program: it provides the principal alternative to magnetic fusion and takes advantages of NNSA investments in the National Ignition Facility. The FY 2004 budget, however, eliminates chamber technology for both MFE (magnetic fusion energy) and IFE." The letter ends, "In summary, FESAC finds the Presidential request for fusion research funding in FY 2004 to be not only meager but also harmfully distorted. It terminates components of the program that are truly essential. Fusion research has accepted new challenges and identified new priorities, consistent with the President's stated agenda; fusion scientists want to get on with the job. What is needed is a funding allocation that respects the magnitude and nature of the task at hand."

The University Fusion Association (UFA) sent a letter to the Subcommittee on Energy and Water of the House Appropriations Committee, urging them to add $25 million to the President's FY 2004 budget request. The letter stated, "Without additional resources, carrying out the necessary preparations for ITER in FY 04 with the present budget request of $257 million (unchanged from the FY 03 request) will result in destroying critical elements of the base science and technology part of the fusion program." The letter notes, "In FY 04 and in subsequent years, as we move forward with the ITER project, the necessary additional funding must be provided to ensure that there is a strong U.S. fusion program to participate in and make use of the advances we achieve in ITER." The Congress ultimately added $6.8 million to the President's request and specified that it was to be used for "non-ITER-related activities in the domestic program." The Congress also reduced the DOE-proposed ITER contribution from $12 million to $8 million.

That the US Administration was not interested in the 35-Year Plan, or any plan to produce fusion energy, was made clear in a May 5, 2003, presentation to the National Academies BPAC by Patrick Looney, Assistant Director for Physical Science and Engineering at the White House Office of Science and Technology Policy (OSTP). He told the committee, "There is no agreed upon fusion energy development timeline." Although acknowledging that, in rejoining ITER, President Bush had said, "The results of ITER will advance the effort to produce clean, safe, reliable and commercially-available fusion energy by the middle of this century," Looney said there were "large error bars on the President's estimate and did not constitute a timeline commitment." Furthermore, Looney said, "This is energy science, not (underline not) an energy technology." He said the US decision to join ITER negotiations is not part of a broader fusion initiative. He said, "The ITER decision will not imply endorsement of other fusion-related initiatives." He added, "As (ITER) construction does not begin until 2006, the (ITER) decision will be overall budget neutral until FY 2006." "If the U.S. joins ITER, it would not be as a lead player," he said, and "the U.S. is absolutely neutral on the issue of site." "The U.S. has no interest in hosting ITER," he said. On the positive side, Looney said, "ITER provides U.S. scientists access to the world's most sophisticated burning plasma experiment."

On the other hand, Energy Secretary Spencer Abraham, speaking at a July 29, 2003, hearing of the Senate Energy and Natural Resources Committee, said, "Perhaps in 20–30 years my successor can come before this Committee and explain how the investments we made today have ultimately paid off. What might that Secretary of Energy say? I would hope he or she could say that after successful completion of the ITER experiment, we are now ready to consider construction of a demonstration fusion power plant to deliver electric power to the grid."

Report of the Burning Plasma Assessment Committee

In the fall of 2003, the National Academies BPAC delivered its final report, entitled "Burning Plasma: Bringing a Star to Earth." The 170-page report concluded that "a burning plasma experiment is critically needed to advance fusion science," but that "undertaking a burning plasma experiment cannot be done on a flat budget." A few months later, the President sent his FY 2005 budget request to Congress, asking for the same amount ($264 million) that Congress had appropriated in FY 2004. Within that flat budget, however, the DOE proposed providing $38 million to ITER (an increase of $30 million over FY 2004) and, once again, stated its intent to terminate all fusion technology programs. The budget document also noted the intent of the Office of Fusion Energy Sciences (OFES) to redirect its inertial fusion energy (IFE) from IFE technologies to support of science-oriented studies of High-Energy-Density Laboratory Physics (HEDLP).

The BPAC recommended that ITER be incorporated as an integral element of the US fusion program and not considered as separable from the US domestic program. They said, "A strategically balanced U.S. fusion program should be developed that includes U.S. participation in ITER, a strong domestic fusion science and technology portfolio, an integrated theory and simulation program, and support for plasma science." They recommended that the USA should participate in ITER but, "if the negotiations fail, the United States should continue, as soon as possible, to pursue the goal of conducting a burning plasma experiment (nationally) with international partners." They commented, "As the ITER project develops, a substantial augmentation in fusion science program funding will be required in addition to the direct financial commitment to ITER construction."

DOE Office of Science Director Ray Orbach then asked the FESAC to "identify the major science and technology issues that need to be addressed, recommend how to organize campaigns to address those issues, and recommend the priority order for these campaigns (through 2014)." FESAC member Charles Baker (University of California, San Diego) was asked to chair the FESAC subpanel, with a report due in July 2004. Stewart Prager (U. Wisconsin) was vice-chair. Orbach indicated that the FESAC could assume "that funding required for ITER construction is provided in addition" to funds required by the domestic program. However, after the President's FY 2005 budget came out in early 2004, it was clear that ITER funding was in fact beginning to come out of the domestic program. In addition, Orbach made clear that he viewed fusion as a science and not an energy program. In a letter to John Lindl, an inertial fusion scientist at the Lawrence Livermore National Laboratory, Orbach said, describing the cuts to the fusion technology programs, "The issue really is the degree to which our Fusion Energy Sciences program should become an energy development program. The Administration position on this issue is that now is not the right time for us to invest in energy related R&D for fusion, either for MFE or IFE." Orbach said, "For MFE, funding for energy relevant technology R&D will wait for the results of ITER. Similarly, for IFE, we will wait for the achievement of ignition and gain before investing in the technology required for energy applications." "Until we are confident that we understand the science of fusion, we would be undertaking an unacceptable risk to commit the sums required to develop the technology needed to apply that science," Orbach said.

Major Advances: Z, OMEGA, LHD

In spring 2003, scientists at Sandia National Laboratories succeeded in compressing a small capsule containing fusion fuel using X-rays from the Z pulsed power facility. Although such compression experiments had been successfully performed previously using X-rays produced by lasers inside a hohlraum, this was the first time such experiments had succeeded using X-rays produced from the Z pulsed power facility. Z pinches have the attractive feature of being relatively low in cost. Concepts for such repetitively pulsed fusion power plants were also conceived and evaluated. A several-year study led by Craig Olson resulted in designs of high-yield inertial fusion power plants pulsing once every 10 s. Two types of target designs are shown in Fig. 9.1.

In April 2004, scientists at the University of Rochester Laboratory for Laser Energetics (LLE) began performing implosion experiments on deuterium-filled cryogenic targets using the OMEGA laser. They successfully produced targets with cryolayer smoothness close to 2 μm rms and produced record fusion neutron yields in close agreement with the predictions of one-dimensional hydrodynamic codes. During FY 2004, a total of 1,558 target shots were taken on the OMEGA Laser Facility. This was a record high number of shots ever taken in a single year by OMEGA or any comparable-sized inertial fusion facility.

In early 2005, scientists at the National Institute for Fusion Science in Japan sustained a 20-million (2 keV) degree temperature plasma for 30 min in the billion-dollar-class Large Helical Device (a stellarator). During the 30-min run, 1.3 billion joules of energy was poured into the plasma, setting a new world record in this regard. As part of the operation, they demonstrated the application of sophisticated radio-frequency power technologies to heat and sustain the plasma. Approximately 700 kW of power was continuously applied at the electron and ion cyclotron frequencies.

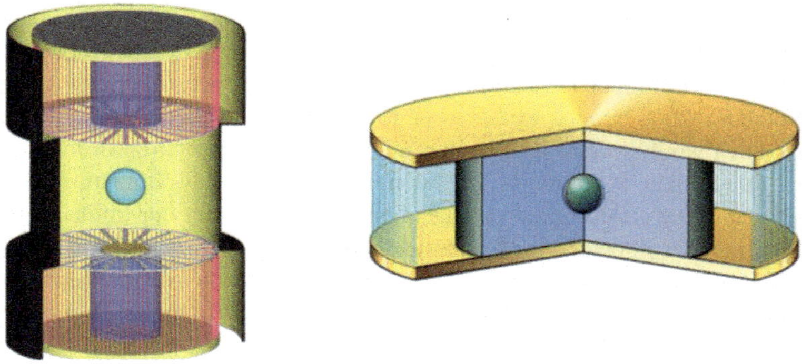

Fig. 9.1 Two types of inertial fusion targets being explored at Sandia National Laboratories

A FESAC panel charged with reviewing the IFE program, chaired by Rulon Linford (University of California), reported in mid-2004 [91]. Contrary to Dr. Orbach's view of doing the science and technology for fusion in series, the panel urged the DOE to carry out "a coordinated program with some level of research on all the key components (targets, drivers and chambers), always keeping the end product and its explicit requirements in mind." The wording was intended to call attention to the fact that DOE had announced termination of all its efforts on targets and chambers for IFE as part of its plan to end all work on fusion technology.

The Congress added $12 million to the President's FY 2005 budget request, directing it to the domestic program and directing "the Department to reduce its planned expenditures on ITER." Meanwhile, the DOE had reestimated the cost of its US ITER contribution, increasing it from $500 million to $1.12 billion.

ITER vs. the US Domestic Fusion Program

FY 2006 was to be the first year of ITER construction funding. The President's budget requested a $17 million increase in the total OFES budget. However, the budget requested a $51 million increase for ITER, thereby necessitating the proposal of $34 million in cuts to the ongoing domestic fusion program. The DOE had previously eliminated the fusion nuclear technology program during FY 2004 and 2005. The additional reductions were to be found by eliminating the fusion materials research program ($7.3 M), halving the effort on heavy-ion inertial fusion ($7.2 M), reductions in ongoing tokamak experiments and theory ($7.4 M), reductions in non-tokamak (alternate concept) research ($10 M), and reductions in enabling plasma technology ($3.0 M).

Testifying on the FY 2006 budget request at a hearing of the House Science Committee Subcommittee on Energy, DOE Office of Science Director Ray Orbach said, "In the FY 2006 budget, we have had to reduce somewhat the domestic (fusion) program, but I would like to look at that in terms of a reorientation of the domestic program rather than a reduction." Subcommittee chair Judy Biggert (R-IL) told Orbach, "I am very concerned about the significant amount of our limited resources that this budget has allocated to the international fusion experiment known as ITER, which doesn't even have a home yet." She added, "I must again express skepticism and concern about the moving target that is the U.S contribution to the ITER project."

In an April 11, 2005, letter to Orbach, the FESAC said, "FESAC is deeply troubled by the President's proposed budget for FY 2006 and its implications for later years. In particular, the core (fusion) program cannot shoulder a significant portion of the ITER construction costs without dismantling the fusion scientific enterprise." FESAC was transmitting the results of its Priorities and Balance Panel, chaired by Charlie Baker, mentioned earlier. FESAC reminded Orbach that the charge letter they had received from him requesting that study had instructed them to assume that funding for ITER construction would be provided above the current level of the US domestic program.

In the fall of 2005, Congress provided the $17 M increase for fusion requested by the President for FY 2006, but declined to authorize the cuts requested in the domestic program. Instead of allowing the DOE to cut the domestic fusion program by $34 M, the Congress directed that $30 M be retained in the domestic program and that proposed ITER funding be reduced accordingly. In the conference report, the House-Senate appropriators said, "As in previous years, the conferees direct the Department to fund the US share of ITER in year 2007 through additional resources rather than through reductions to domestic fusion research or to other Office of Science programs."

On June 28, 2005, representatives of the governments of the European Union, Japan, Russian Federation, People's Republic of China, Republic of Korea, and the United States met in Moscow and signed a Joint Declaration designating Cadarache, France, as the ITER site. The group ratified an agreement on cost sharing among the Parties. The construction phase of ITER was to begin in 2006, with first plasma operation in 2016. First full-power (500 MW thermal) deuterium–tritium operations were to begin in 2021. The cost of ITER was still advertised as $5 billion (in year 2000 dollars), though the DOE advised Congress that the US 10% share would cost $1.122 billion. DOE's Orbach said that number would be a "cap" on the US expenditure. In its FY 2007 budget request, DOE provided Congress with an ITER funding profile for the US contribution as follows:

FY 2006	$19.3 M	FY 2011	$181.3 M
FY 2007	$60.0 M	FY 2012	$130.0 M
FY 2008	$160.0 M	FY 2013	$116.9 M
FY 2009	$214.5 M	FY 2014	$30.0 M
FY 2010	$210.0 M	Total	$1,122.0 M

Energy Policy Act of 2005

In late July 2005, the Congress passed and President Bush signed into law the Energy Policy Act of 2005. A section in the Act on fusion policy states, "It shall be the policy of the United States to conduct research, development, demonstration, and commercial applications to provide for the scientific, engineering, and commercial infrastructure necessary to ensure that the United States is competitive with other countries in providing fusion energy for its own needs and the needs of

Major Advances: NIF, HIF, OMEGA, EAST

On December 1, 2005, LLNL issued a press release announcing a "round of successful laser experiments at the National Ignition Facility (NIF), validating key computer simulations and theoretical projections relevant to the plasma and X-ray environment necessary to achieve ignition." Using 8 of the planned 192 laser beams, scientists irradiated a series of different-sized hohlraums with 2 ns laser pulses of about 8 trillion watts, which produced radiation temperatures in agreement with both theoretical expectations and previous experience from other facilities. The results were published in the November 18, 2005, issue of Physical Review Letters. Ed Moses led the NIF program at LLNL.

During 2006, scientists at the Lawrence Berkeley and Livermore National Laboratories working in heavy-ion fusion (HIF) program demonstrated accelerator injector technology with potentially significant economic and technical impact on the size and cost of injector systems for heavy-ion-fusion inertial fusion drivers. An ion beam for heavy-ion fusion must have high "brightness," that is, high current and low emittance temperature, in order to deliver high power onto a small target spot. A first step in achieving this is to begin with the minimum possible emittance at injection. The scientists injected 119 beamlets into an electrostatic quadrupole channel, demonstrating a merging of the beamlets. Grant Logan led the HIF program.

Also during 2006, scientists at the University of Rochester Laboratory for Laser Energetics performed scaled experiments on the OMEGA Laser Facility to validate the design of ignition targets for inertial confinement fusion ignition experiments on NIF. The experiments showed good agreement with scaling models. Bob McCrory led the Rochester effort.

On September 26, 2006, scientists in China began experiments on their new superconducting tokamak, called EAST. The project is the result of an international collaboration among the Institute of Plasma Physics in Hefei, China; General Atomics; and PPPL. Chinese officials stated the EAST "will be a unique facility to explore some critical issues relating to steady-state operation with shaped plasma cross-section in the next few years." EAST is designed to produce plasmas lasting for about 1,000 s.

other countries, including by demonstrating electric power or hydrogen production for the United States energy grid using fusion energy at the earliest date."

The Act called on the Secretary of Energy "to negotiate an agreement for United States participation in ITER," but specifies that "no federal funds shall be expended for the construction of ITER until the Secretary has submitted to Congress . . . a report describing how United States participation in the ITER will be funded without reducing funding for other programs in the Office of Science (including other fusion programs), and 60 days have elapsed since that submission."

The Act says, "fusion science, technology, theory, advanced computation, modeling, and simulation" should be strengthened and that "new magnetic and inertial fusion research and development facilities (should be) selected based on scientific innovation and cost effectiveness, and the potential of the facilities to advance the goal of practical fusion energy at the earliest date practicable."

More Proposed US Domestic Fusion Budget Cuts

In early 2006, the President sent his FY 2007 budget request to Congress, requesting $319 million for OFES (an increase of $38 million over FY 2006). In it he requested $60 million for ITER and proposed once again to cut funding for heavy-ion fusion and high-energy-density physics (by $4 M), innovative confinement physics (by $2 M), and fusion materials research (by $2.4 M). In response, the University Fusion Association (UFA) wrote, "The Administration's proposed budget for FY 2007 would make broad cuts in university research programs, seriously damaging U.S. capabilities to benefit ITER."

In early 2007, President Bush sent to Congress an aggressive FY 2008 fusion budget request for $428 million, compared to $319 million in FY 2007. He requested the full $160 M needed for ITER construction, along with a slight ($9 million) increase for the domestic fusion effort ($268 million compared to $259 million in FY 2007). Congress eventually rejected the large increase for ITER, providing OFES with only $290 million in FY 2008. This threw the US planned ITER funding profile into disarray. In early 2008, DOE scrapped its previous ITER funding profile and told Congress its new projected contribution to the ITER project would total somewhere between $1.4 and $2.2 billion. Nevertheless, in early 2008 the President asked Congress to provide OFES with $495 million, including a $214.5 million contribution to ITER (the amount contained in the old funding profile). Congress would eventually provide $395 million to OFES, a $100 million increase, all for ITER.

Davies, Roberts, and Willis Retire

Dr. N. Anne Davies, head of the US fusion program since 1989, and Dr. Michael Roberts, Director of the ITER and International Division under Davies, both retired from federal service on April 3, 2006. Their retirements followed on the earlier retirement (on April 1, 2005) of Dr. John Willis, Director of the Research Division under Davies, after 34 years of federal service.

The retirements left the DOE Office of Fusion Energy Sciences (OFES) without its top-level management team. Orbach's deputy, Dr. James Decker, assumed Davies position on an "acting" basis, while Davies successor was recruited. Eventually (on March 1, 2007) Dr. Raymond Fonck, a plasma scientist from the University of Wisconsin, assumed the post. He would stay in the position for only 17 months, however, before returning to the university. Fonck had been the recipient of Fusion Power Associates 2004 Leadership Award.

Fusion Power Associates presented its 2007 Distinguished Career Award to Dr. Davies and, later, presented Special Awards to Roberts and Willis.

2007 Inertial Fusion Energy Workshop

An Inertial Fusion Energy Science and Technology Workshop was held on April 24–27, 2007, in San Ramon, California [92]. The workshop focused on the science, technology, and policies needed to make inertial fusion an attractive energy source. Four breakout sessions were held, chaired by Grant Logan (LBNL), Steve Obenschain (NRL), Craig Olson (SNL), and myself (FPA). On the last day, I presented a summary of the four breakout sessions as follows:

- IFE, with the exception of some aspects of heavy-ion fusion, has been funded by Congressional add-ons (the HAPL program) and Laboratory Directed Research and Development (LDRD). NNSA (the primary funding source for inertial fusion) is focused on single-shot facilities and technologies; OFES is focused on fusion "science" and not energy development. A "home" is needed in DOE that is interested in fusion energy development that would be interested in receiving and implementing an IFE strategic plan. This will require a change in policy in the Executive Branch. Previous fusion community and DOE review panel studies all have shown that MFE and IFE are equally credible as fusion energy approaches; yet current DOE policy, whether deliberate or not, has the effect of holding back IFE development.
- A defining event for triggering a breakout for implementation of an IFE plan is the anticipated achievement of ignition on NIF in the 2010–2012 time frame. The DOE should be prepared to capitalize on that success. Therefore, specific programs should be carried out in the immediate future in order to be prepared to respond to that event. These include rep-rated driver, target and chamber development, computation, and systems' studies for IFE. The funding required for these efforts, pre-NIF ignition, is approximately \$60–70 M per year, which is modest compared to the currently funded inertial confinement fusion (ICF) efforts within NNSA. In recent years, funding for IFE-specific research (\$25–30 M per year) has been entirely due to year-to-year Congressional actions. DOE has no visible plans to support such efforts. "Energy-related" research in the new

HEDLP (High-Energy-Density Laboratory Physics) program would be complementary to IFE but is no substitute for a focused IFE effort.

- The cornerstone of an IFE breakout strategy is a program aimed at construction of a rep-rated IFE fusion test facility. The detailed configuration of this facility is not fixed yet, but it is envisaged as the last major facility prior to a Demo (demonstration power plant). As part of a program leading to construction, subsystem prototypes and component development would be required in order to finalize design and proceed to construction. These include driver development for lasers, heavy ions, and Z-pinch drivers, target design and fabrication, chamber technologies, and detailed conceptual design. Modularity and separability of subsystems lend themselves to a cost-effective development path as part of a fusion test facility program.
- The readiness to proceed with a fusion test facility, post-NIF ignition, depends heavily on obtaining a stable source of funding. Currently, funding for all aspects of IFE is in doubt every year.
- As part of the IFE strategic plan, nontechnical questions should be addressed that answer issues that most likely would be raised by nontechnical people, such as why is IFE a desirable product for society.
- The plan should contain items such as vision, detailed R&D tasks, budget, and schedule. The plan should stress the cost-effectiveness of IFE development in light of the substantial science and technology base being developed by NNSA. It should also describe the potential for a faster development path relative to some previous fusion plans. Since replacement of existing power plants and construction of new ones will likely begin in earnest around 2050, fusion should attempt to put forth a plan that would have a commercial power plant ready for deployment around that time, if possible.
- The IFE strategic plan should include partnerships among labs, universities, and industry. Industry involvement in the development will make it more likely that the IFE program will lead to a commercial product. Power plant studies are also needed to provide an attractive long-range vision for IFE and to highlight areas in which improvements are needed.
- While the US IFE community should collaborate with the growing international IFE effort, the plan should be capable of being implemented domestically and focused on a product that would be attractive to the US market.
- The general features of IFE described in the Linford FESAC IFE panel report [91] are still current. However, much progress has been made since then. For example, more serious efforts are underway in the USA and elsewhere on fast ignition, more attractive designs and techniques for irradiating targets have been proposed, and a shortened development path has been suggested.
- The current DOE plan to emphasize and expand efforts in HEDLP will strengthen the physics basis for IFE. It is, however, not a substitute for the more urgent need for energy-related and IFE-focused efforts on IFE development.

The NNSA inertial confinement fusion (ICF) budget held up well during the decade of the 2000s, starting at $475 M in FY 2000 and ending at $479 in FY 2008, with a peak of $544 in FY 2006. Variations in ICF funding were due largely to the needs of NIF construction, with the exception that Congress consistently added about $25 M each year for the IFE-oriented HAPL program (although NNSA consistently did not ask for the HAPL funds each following year).

Fifty Years of International Fusion Collaboration

In October 2008, the United Nations International Atomic Energy Agency (IAEA) held its 22nd Fusion Energy Conference in Geneva, the site of the 1958 Geneva Conference on the Peaceful Uses of Atomic Energy where fusion research was first declassified, to celebrate the 50-year anniversary of IAEA conferences on fusion international collaborations. Dale Meade (PPPL) presented a paper, "Some Highlights of 50 Years of Fusion Research," summarizing technical progress [93]. Sadly, on July 19, just a few months before the conference, British fusion pioneer Nicol Peacock, who had been instrumental in starting the world tokamak "bandwagon," passed away. He had been a leader of the team that went to Moscow in 1969 to measure the plasma temperature in the T-3 tokamak [17].

The Obama Administration: 2009–2012 10

A ship without a rudder may wander aimlessly among perilous isles, yet sink not to the bottom.

–Kahil Gibran
Author, The Prophet

Washington is ruled by one principle: there is no money.

–Paul Gilman
Aide to Sen. Peter V. Domenici

In November 2008, Barack Obama (a Democrat) won election to the US Presidency, and both the House and Senate remained firmly under control of the Democratic Party. Scientists, in general, and fusion researchers, in particular, were enthusiastic as new presidential appointments were announced for the White House (OSTP and OMB) and the Department of Energy.

New Appointments

As Science Advisor and head of the White House Office of Science and Technology Policy (OSTP), the President chose John P. Holdren from Harvard University. Holdren had extensive experience with fusion issues, dating back to the early 1970s, when he published a paper "Analytic Approximation to Collisional Distribution in Mirror Plasmas" while working in the magnetic fusion energy program at the Lawrence Livermore National Laboratory. In the 1970s and 1980s, he also wrote several papers on environmental and safety issues associated with fusion power plants. In 1989, he chaired the Senior Committee on Environmental, Safety, and Economic Aspects of Magnetic Fusion Energy (ESECOM). That report [79] stated that the "advantages (of fusion) are potentially large enough to make a difference in public acceptability of magnetic fusion energy, as compared

S.O. Dean, *Search for the Ultimate Energy Source: A History of the U.S. Fusion Energy Program*, Green Energy and Technology, DOI 10.1007/978-1-4614-6037-4_10,

to fission." As a member of the President's Council of Advisors on Science and Technology (PCAST) from 1994 to 2001 under President Clinton, he chaired two studies on fusion policy, as described in Chap. 8 [81, 82]. In the 1997 PCAST report, he recommended a $100 million increase in the fusion budget which then (FY 1997) was $232 million.

For Secretary of Energy, President Obama chose Steven Chu, a 1997 Nobel Laureate for his atomic physics research using lasers. He worked at Bell Labs and Stanford University before becoming Director of the Lawrence Berkeley National Laboratory (LBNL) in 2003. As LBNL Director, Chu had the Heavy Ion Fusion IFE program among his responsibilities although, it is said, he paid scant attention to it. In a speech at the National Press Club on September 17, 2008, Chu said "Our country needs to act quickly with fiscal and regulatory policies to ensure widespread deployment of effective technologies that maximize energy efficiency and minimize carbon emission." He noted, "However, most companies are reluctant to invest in research into transformational technologies that may not see commercialization for 10 years, even though such technologies could dramatically change the entire energy landscape." Though he did not mention fusion (or other) technologies specifically, he said, "I believe that aggressive support of energy science and technology, coupled with incentives that accelerate development and deployment of innovative solutions, can transform the entire landscape of energy demand and supply. It is especially important that we invest in those ideas that industry has not yet been willing to explore."

For Undersecretary for Science, President Obama chose Steven Koonin, chief scientist since 2004 at British Petroleum, the world's second largest oil company. Koonin had previously been at the California Institute of Technology, where he also served as Provost from 1995 to 2004. Koonin had a familiarity with inertial confinement fusion dating over several decades, having served on and chaired numerous inertial fusion review and advisory committees. He chaired a 1990 National Academy of Sciences inertial confinement fusion program review, for example, and also served in the mid-1990s on DOE's Inertial Confinement Fusion Advisory Committee (ICFAC). Among the Academy committee's recommendations was that the DOE should accelerate the construction of the krypton fluoride Nike laser at the US Naval Research Laboratory and that the construction of the OMEGA laser at the University of Rochester should "be started immediately." He was a recipient of Fusion Power Associates 1994 Leadership Award in recognition of the "major impact he has had on the programs and plans of the inertial confinement fusion program."

William Brinkman, a Senior Research Professor in the Physics Department at Princeton University, was chosen as Director of the DOE Office of Science to succeed Ray Orbach, who had been a champion for the ITER project. Brinkman had joined Bell Laboratories in 1966, became the Director of the Physics Research Laboratory there in 1981, and eventually became Research Vice President in 2000. Brinkman was familiar with the major fusion effort at the Princeton Plasma Physics Laboratory and had for a time (1984–1987) also been Vice President of Research at Sandia National Laboratories, where a major inertial confinement fusion effort was in progress.

As a result of these appointments, the US fusion community felt that individuals were now in place who understood our science and our goals. In short, these were "people we could talk to."

Inertial Fusion Energy

Shortly after the Senate confirmed him in 2009, I met with Koonin in his DOE office to discuss the US fusion program in general and the inertial fusion program in particular. The civilian inertial fusion energy (IFE) effort was not having much luck in finding a "home" in the DOE. The Office of Fusion Energy Sciences (OFES) budget was consumed by the needs of magnetic fusion energy (MFE), particularly tokamak research and ITER. Managers of the NNSA weapons budget, which supported the inertial confinement fusion (ICF) program, maintained that the energy aspect of ICF was not in their "mission." Therefore, they refused to request funds for IFE-related research, even though Congress had provided them funds for the IFE High Average Power Laser (HAPL) program for over a decade.

I provided Koonin with a "white paper" I had prepared with help from several members of the inertial fusion community [94], entitled "The Rationale for an Expanded Inertial Fusion Energy Program." The technical basis for the Rationale was primarily that the NNSA ICF program, via the nearing-completion National Ignition Facility (NIF), would soon be demonstrating ignition of fusion capsules in single shot mode and, consequently, "enhanced funding for IFE R&D is needed in the near term to prepare to expeditiously proceed beyond NIF to the energy application of inertial fusion." We discussed the problem that neither OFES nor NNSA seemed willing or able to take on the IFE mission. Koonin understood the problem and bemoaned finding what he called a "stovepipe" mentality within DOE. He said he was going to work on breaking down such barriers to accomplishment of overall DOE goals. He felt he needed some kind of outside review to assist him in addressing the IFE issue, however. We discussed various possible ways of establishing a review, including FESAC, JASONS, SEAB, and the National Academies. I was somewhat disappointed that he did not feel he could just "take the bull by the horns" based on his own knowledge and judgment. Any review would take time, the Academies the longest time but perhaps the most influential. He chose to set up an Academy review which, 3 years later (early 2012), had produced only an interim report [95].

About the same time I was meeting with Koonin, Physics Today editor and science writer David Kramer conducted an interview with Koonin and asked him "What about Fusion?" Koonin's response, as published in the September 2009 issue of Physics Today, was "Speaking for myself, I'd like to see more focus and drive in our fusion programs than we've had in the past. ITER is going to turn on in perhaps 2020-something, in terms of (achieving) a burning plasma. That's too long. We should be looking for ways to accelerate that, or perhaps exploring other methods to get to fusion energy. Inertial fusion energy is another option that the department is pursuing through the NNSA (National Nuclear Security Administration) and at the National Ignition Facility. I would keep an eye on NIF as perhaps offering an alternative route to fusion energy."

While I was meeting with Koonin, William Brinkman, who had not yet been confirmed as Director of the DOE Office of Science by the Senate, walked in. His first comment/question to me was "What about ITER?" He said he was concerned by its rising cost. I said that I was also concerned that a fusion engineering test reactor should be so costly to construct but that I thought it was a "done deal" and that the government had committed to it. I said I thought it was possible that ITER would collapse under its own (financial) weight, but that I would hate to see the USA be the first to pull out and be blamed for its collapse. Brinkman was later confirmed, became a member of the ITER Council, and also became an advocate for continued US participation in the project.

NIF Begins Operation and Looks Beyond Ignition

The National Ignition Facility (NIF), which had been under construction at the Lawrence Livermore National Laboratory in California, fired its first 192 laser beam full power shots into a test cell in spring 2009. The lasers delivered almost 2 million joules of infrared light, which was then successfully converted to 1.1 million joules of ultraviolet light (as is needed for fusion ignition tests). The energy and pulse shape (in time) of the beams approximately matched that predicted to be needed for achieving ignition. In June 2009, the NIF laser system began firing all 192 beams onto targets at lower energy in order to characterize the X-ray drive achieved inside the small cylinders (hohlraums). These initial target experiments were conducted on smaller capsules than would be used for the ignition experiments. In early 2010, the NIF laser put 1 MJ onto a target. These targets used gas-filled capsules that act as substitutes for the fusion fuel capsules that were used later. NIF Director Ed Moses said, "This is an incredible milestone on our journey to ignition. We are well on our way to achieving what we set out to do—controlled, sustained nuclear fusion and energy gain for the first time ever in a laboratory setting."

On September 29, 2010, NIF scientists performed the first integrated ignition experiments, firing all 192 laser beams and 1 MJ onto a cryogenically layered capsule containing a mixture of tritium, deuterium, and hydrogen. All systems operated successfully. Project Director Ed Moses said, "This is a great moment in the 50-year history of Inertial Confinement Fusion." Ignition was not achieved, however, and fusion ignition tests are continuing.

As NIF neared completion, LLNL began a study aimed at identifying a path to a commercial fusion power plant. Dubbed "LIFE" (for Laser Inertial Fusion Energy), LLNL envisaged a very rapid post-NIF development program to commercialization [96].

Funding Improvement and Management Changes

The Congress provided OFES with $403 million for FY 2009, including $124 million for ITER and slightly over $100 million above the FY 2008 allocation. For FY 2010, President Obama requested, and the Congress later provided, $426 million. The OFES program also received $91 million in "stimulus" funding in FY 2009, as part of the nation's effort to stimulate the US economy in response to the economic recession that had been under way in the country for about 2 years. For the ICF program in NNSA, the Congress provided $442 million in FY 2009 and $458 million in FY 2010. The NIF funding within NNSA had never been threatened financially like ITER was within OFES. The NNSA ICF budget had been above $400 million since FY 1998.

Effective April 1, 2009, Stewart Prager became director of the Princeton Plasma Physics Laboratory (PPPL), succeeding Rob Goldston. Prager was previously a professor and fusion scientist at the University of Wisconsin and a former chairman of FESAC.

Effective June 7, 2009, Ed Synakowski became head of OFES, replacing Ray Fonck who had resigned the post in August 2008. Synakowski was a physicist who had worked on TFTR and NSTX at PPPL after receiving his Ph.D. from the University of Texas in 1988. He had been head of the LLNL fusion energy program since 2006.

In early 2010, President Obama nominated another fusion veteran to an important post. Donald L. Cook was named Deputy Administrator for Defense Programs at NNSA. He had worked on a variety of programs, including inertial fusion, at the Sandia National Laboratories beginning in 1977 and had managed the ICF and pulsed-power programs there in 1984–1999. He was a recipient of Fusion Power Associates 1993 Leadership Award.

President Obama's appointments, while having extensive knowledge and interest in fusion, nevertheless had other demands on their time and many other responsibilities and interests at the DOE. Their jobs were made even more challenging by the fact that a great economic recession had begun and was continuing in the USA and in the world. The US government engaged in large amounts of deficit spending, both to prop up the domestic economy and to engage in wars in Iraq and Afghanistan (over $1 trillion for the wars). Some of the "stimulus" funds ($91 million) found its way to the fusion program, resulting in construction of a small heavy-ion accelerator at the Lawrence Berkeley National Laboratory, aimed at studying high-energy-density physics related to inertial fusion.

In early 2010, President Obama submitted his FY 2011 budget to Congress. It proposed only $380 million for fusion (OFES) compared to $426 million appropriated in FY 2010. The budget request stated that spending for ITER would be reduced by $55 million ($80 million, compared to $135 million in FY 2010), resulting in a proposed $9 million increase in the domestic fusion program. The decrease in ITER funding reflected a recognition that the ITER project was proceeding more slowly than anticipated, resulting in schedule slippages. The inertial confinement fusion program in NNSA would receive $482 million, compared to $437 million in FY 2010, mostly for "NIF diagnostics, cryogenics, and experimental support." The OFES program eventually received only $367 million, including $80 million for ITER.

ITER Changes

During 2008 and 2009, international reviews were under way on the management, schedule, and cost of the ITER project. Originally, ITER construction completion was scheduled for 2016. But in June 2008, the ITER Council approved a 2-year slip in first operations to 2018, with first fusion (DT) plasmas in 2023. At a Council meeting in June 2009, the Council kept (temporarily) the 2018 "first plasma" date but slipped the schedule for DT operations to 2026. Further schedule slippage was considered at a November 2009 Council meeting, but that decision was deferred. These issues (management, schedule, and cost) came to a head during the first 6 months of 2010.

At a February 23–24, 2010 meeting of ITER Heads of Delegations (China, EU, India, Japan, Korea, Russia, and the USA), the ITER Organization proposed an additional schedule slippage of a year (to November 2019 for first plasma). ITER design changes were partly to blame for the schedule slippages, but estimated costs to complete the project were also exceeding the planned expenditures of some of the Parties, most notably the EU and USA. In the USA, for example, the cost of its ITER contribution had been stated in 2006 as $1.12 billion, but in 2008 the DOE revised this to a "range" between $1.4 and $2.2 billion. In 2010, the EU acknowledged that it had revised the estimate for its 45% share upward by more than a factor of two. The European Union warned the European Parliament and the European Council to expect that Europe's share of ITER construction could increase from Euro 2.7 billion to Euro 7.2 billion. In addition, US Head of Delegation, William Brinkman (Director of the DOE Office of Science) told a March 8, 2010 FESAC meeting that the Heads of Delegations had received a "very negative" report on the ITER management structure. Brinkman told the FESAC, "If I could get my hands on the person who proposed the current management structure, I would strangle him." One of the key management issues was lack of effective oversight by the ITER Organization over the activities of the Domestic Agencies of the ITER Parties.

At its meeting during the last week of July 2010, the ITER Council approved the 1-year schedule slippage (to November 2019) and slipped DT operations another year (to 2027). The Council approved a new "baseline" cost estimate ($19 billion) that was approximately double the previous agreed-upon estimate (or quadruple the original estimate). With a 9% share, the US contribution would cost about $1.8 billion. The Council also named a new ITER Director-General, Osamu Motojima, to replace Kaname Ikeda who had led the project since the international ITER agreement had been signed. Motojima was formerly Director-General of the National Institute for Fusion Science (NIFS) in Japan and had led the construction of the billion-dollar-class Large Helical Device (LHD) there. He was the recipient of Fusion Power Associates 2008 Distinguished Career Award.

ITER continued to face cost and schedule issues into 2012. Projected doubling plus of the EU's ITER cost presented a major problem but was eventually solved (in principle) in late 2011 by an "ad referendum" agreement among the European Council, the European Parliament, and the European Commission on where to find additional funds through 2013. The tsunami in Japan in March 2011 called into question the ability of Japan to maintain their planned contribution to ITER, but this was also eventually solved. The situation in the USA, however, was worsening due to concern in Congress and the Administration over US deficit spending. Pressure was mounting to hold level or reduce so-called "discretionary" spending. Fusion fit in this category. As a result, President Obama's FY 2013 budget request, presented to Congress in early 2012, held the US total fusion effort approximately level while also providing ITER approximately $100 million less than needed to meet the US commitment to the ITER schedule. ITER was proposed for a $45 million increase over its FY 2012 level ($150M compared to $105M), compensated by a proposed similar reduction in the US domestic fusion effort. Although DOE did not publicly acknowledge that the total cost of its ITER share was continuing to escalate, it was common knowledge in the US fusion community that it was approaching $2.5 billion.

Magnetic Fusion Concepts Narrowing

A "conflict" has always existed within the magnetic fusion program on the issue of how many "concepts" to pursue and to what extent "alternate concepts" (to the front-runner tokamak) should be supported. After the mid-1980s (FY 1986,87,88) budget cuts, which led to the demise of the main alternate concept (magnetic mirrors), scientists working on smaller alternates (pinch, field-reversed, etc.) were told to justify their efforts in terms of the contributions they might make to tokamak development. Two DOE fusion advisory panels in the 1990s asserted that this was inappropriate and that an alternate concepts program should be maintained for its own sake [97, 98]. During the 2000s, OFES changed the title of this budget category to "Innovative Concepts" and included tokamak-related work in this category, including the NSTX tokamak and stellarators. This had the effect of making the activity on "real alternatives" to the tokamak appear larger than in fact it was. Starting in 2003, as the commitment to ITER took shape, pressures mounted to focus even more on providing support for the tokamak concept. Furthermore, pressures from OMB and the Office of Science continued to urge the magnetic fusion program to describe itself and its goals in "scientific terms" rather than in terms of fusion "concepts." A series of workshops and FESAC studies, called ReNew, laid out the fusion effort in terms of scientific issues to be studied (see FESAC report postings at http://science.doe.gov.fes/).

When Ed Synakowski took over as head of OFES in June 2009, he not only bent to these pressures to narrow the fusion program, he embraced them. Furthermore, as it became clear to him that there was no indication that he would be able to secure funding for a major new facility in the USA, he became convinced that the future of the US magnetic fusion effort lay with support for ITER and told fusion researchers to plan on working on newer or more capable tokamak facilities overseas. This "strategy" was very unpopular (to say the least) in the US fusion community.

The first concrete indication of a change in fusion policy under Synakowski came in the form of a solicitation for proposals in the Innovative Confinement Concepts (ICC) program, issued by OFES in early 2010. The solicitation described the ICC program as one that "explores improved pathways to practical fusion power by addressing critical problems that hinder the tokamak concept, such as plasma disruption, heat load on internal components, and operational and maintenance complexity." The solicitation said, "Compared to previous ICC solicitations, an increased emphasis will be placed on those proposals that have the most promise of making such linkages and addressing the (tokamak) issues. Overall, support of research that can best help deepen the scientific foundations of understanding and improve the tokamak concept is an important focus area of this Announcement." At a FESAC meeting March 10, 2010, Glen Wurden (Los Alamos National Laboratory) said the solicitation "substantially changes the nature and intent of the ICC program." He said, "This was certainly not what the ICC program did in the past." In response to a question from FESAC member Rulon Linford, Synakowski confirmed that the wording represented a deliberate intent on his part to put less emphasis on the "confinement concept" aspect of the term "ICC" and more on the "innovative" aspect, in support of the broader scientific aspects of the fusion program. He said that budget realities make it unlikely that any of the alternate concepts would be able to have new, larger facilities in the foreseeable future.

When the President submitted his FY 2013 budget in early 2012, the fusion budget description no longer had a category for Innovative Confinement Concepts (ICC). Five ICC programs were quietly terminated by OFES during 2011, including the Levitated Dipole, a joint MIT-Columbia University experiment based at MIT. Programs previously funded in the ICC category were now included under a category called Experimental Plasma Research, and the proposed budget for the category had been reduced from $18 million to $11 million. In the FY 2013 budget submission, scientists whose programs were aimed at finding an "alternate" to the tokamak concept could no longer find mention of their specific programs or concepts in the budget write-up.

Magneto-Inertial Fusion Approach Gets a Boost

Magneto-inertial fusion (also called magnetized target fusion) is an approach that seeks to access fusion plasma densities intermediate between the relatively low-density plasmas typical of tokamaks and other magnetic fusion energy (MFE) approaches and the very high densities typical of inertial confinement fusion (ICF/IFE). This approach, long advocated and studied by scientists at the Los Alamos National Laboratory, received a degree of validation in experiments at the University of Rochester in 2011 and also from computer simulations at the Sandia National Laboratories (SNL). Experiments are under way at Los Alamos [99, 100] and also in conjunction with the Shiva Star facility at the Air Force Research

Laboratory [101]. Experiments at SNL using the powerful Z pulsed-power facility are planned in 2013.

Using the OMEGA laser facility at the University of Rochester, scientists demonstrated enhancement of the fusion yield from a magnetized laser-driven target implosion [102]. A magnetic field of 80 kG was placed around the target and subsequently trapped and compressed by imploding the conductive plasma. The observed ion temperature and fusion neutron yield were enhanced by 15% and 30%, respectively. Although the magnet coil surrounding the target introduces an asymmetry that would normally interfere with obtaining a spherically symmetric implosion, a spherical compression was nevertheless achieved using the relatively new technique known as polar direct drive.

At Sandia National Laboratories, the results of computer simulations [103] predicted that "scientific breakeven" (fusion energy produced equal to energy deposited in the fusion fuel) could be possible using the 26 megamp current now available in the Z facility if DT were used as the fuel. Experiments on the facility are forecast to be carried out in 2013, but DD will be used as the fuel instead of DT, due to lack of radiation shielding at the facility. The simulation predicts that a larger Z-type facility (with about 3 times the current) could produce energy gains of 1,000 or more. A workshop was held on February 5–8, 2012, to discuss issues associated with this technique [104], which SNL has dubbed MAGLIF (Magnetized Liner Inertial Fusion).

A paper describing the benefits of investigating this intermediate density regime and the technologies for accessing it was described by Irv Lindemuth and presented at Fusion Power Associates December 1–2, 2010 annual meeting [105].

Renewed Interest in Path to a Demonstration Power Plant

A meeting of over 70 experts from fusion centers, industry, and the European Commission was held in May 2011 in Garching, Germany, to mark the start of pre-conceptual design for a fusion power plant. A Department for Power Plant Physics and Technology was set up there under the direction of Gianfranco Federici. Federici said, "We must take on a more system-oriented and integrated approach instead of concentrating on detailed component design." Serge Paidassi, from the European Commission, said "Ideally we would start constructing Demo in 2030 and have it up and running in 2040. This is a very ambitious goal. It would then be possible to implement operation of a commercial power plant by 2050." Paidassi also said, "If you do not involve industry early enough in the process you may develop concepts, ideas, and designs that are not fit for industrial realization." The Thales Group, Siemens, Ansaldo, and Areva were among the industries participating in the meeting.

In the USA, an International Workshop on MFE (magnetic fusion energy) Roadmapping was held at PPPL in September 2011 [106]. Approximately 70 persons from the USA, Germany, India, China, Japan, Korea, France, and Russia attended. The goal of the workshop was "to foster an international technical discussion of the science and technology issues and prerequisites for a Demonstration

Power Plant (Demo), and the missions, requirements, and risk tradeoffs for major nuclear facilities, including ITER, on the path to Demo."

Subsequently, the International Atomic Energy Agency (IAEA) decided to launch a series of annual workshops on planning for a demonstration fusion power plant. The first IAEA Demo Programme Workshop (DPW-2012) was scheduled at the University of California at Los Angeles (UCLA) from 15 to 18 October 2012. The specific objective of DPW-2012 was to discuss a subset of key Demo scientific and technical issues with the aim of defining the facilities and program activities that can lead to their resolution. IAEA stated, "the workshop output will be information that could be used by any party as input to the planning of possible roadmaps to Demo."

Inertial Fusion Energy Review

Soon after I had met with him in the summer of 2009, DOE Undersecretary for Science Steven Koonin announced (in November 2009) that he would set up a review of the prospects for inertial fusion as an energy source. Setting up such a review was not easy or quick, however. It was not until spring 2010 that a statement of task and an agreement had been signed between the DOE and the National Academies of Science and Engineering National Research Council to perform the review, and it was not until December 2010 that a review committee held its first meeting.

The statement of work for the review says, "A committee will be convened to assess the prospects for inertial confinement fusion energy systems. The Committee will prepare a report that will:

- Assess the prospects for generating power using inertial confinement fusion
- Identify scientific and engineering challenges, cost targets, and R&D objectives associated with developing an IFE demonstration plant
- Advise the U.S. Department of Energy on its development of an R&D roadmap aimed at creating a conceptual design for an inertial fusion energy demonstration plant

The Committee will also prepare an interim report to inform FY 2012 budget deliberations. A Panel on Fusion Target Physics will serve as a technical resource to the committee."

The task states, "A Panel on Fusion Target Physics ("the Panel") with access to classified information as well as controlled-restricted unclassified information will serve as a technical resource to the Committee on Inertial Confinement Energy Systems ("the Committee") and will describe, in a report containing only publicly accessible information, the R&D challenges to providing suitable targets on the basis of parameters established and provided by the Committee. The Panel will also assess the current performance of various fusion target technologies."

Although the statement of work called for an interim report in time "to inform the FY 2012 budget deliberations," in fact the interim report [95] (provided to DOE on March 2, 2012) did not even arrive in time to "inform" the FY 2013 deliberations.

Table 10.1 Members of the 2010–2012 National Academies' Committee on IFE

Ronald C. Davidson, Princeton University, *Co-Chair*	David Hammer, Cornell University
Gerald L. Kulcinski, University of Wisconsin, *Co-Chair*	Joseph S. Hezir, EOP Group, Inc.
Charles C. Baker, University of California, San Diego (retired)	Kathryn McCarthy, Idaho National Laboratory
	Lawrence T. Papay, PQR, LLC
Roger O. Bangerter, Lawrence Berkeley National Lab (retired)	Ken Schultz, General Atomics (retired)
	Andrew M. Sessler, Lawrence Berkeley Nat'l Lab.
Riccardo Betti, University of Rochester	John Sheffield, The Univ. of Tennessee, Knoxville
Jan Beyea, Consulting in the Public Interest	Thomas A. Tombrello, Jr., CA Inst. of Technology
Robert L. Byer, Stanford University	Dennis G. Whyte, MIT
Franklin Chang-Diaz, Ad Astra Rocket Co.	Jonathan S. Wurtele, Univ. of California, Berkeley
Steven C. Cowley, UK Atomic Energy Auth.	Rosa Yang, Electric Power Research Institute, Inc.
Richard L. Garwin, IBM Thomas J. Watson Research Center	Malcolm McGeoch, Consultant, PLEX, LLC

Table 10.2 Members of the 2010–2012 National Academies' Panel on ICF Targets

John Ahearne, Sigma Xi, *Chair*	Thomas Mehlhorn, Naval Research Laboratory
Robert Dynes, Univ. of California at San Diego	Merri Wood-Schultz, Consultant
Douglas Eardley, Kavli Inst. for Theoretical Phys	George Zimmerman, Consultant
David Harding, University of Rochester	

When it became obvious that the task would not be completed on the schedule originally envisaged, the statement of task regarding the timing of the interim report was revised to read: "The Committee will also prepare an interim report to inform future year planning by the federal government."

Members of the main review committee and the target physics review panel are shown in Tables 10.1 and 10.2.

The Committee, co-chaired by Ronald Davidson (Princeton University) and Gerald Kulcinski (University of Wisconsin), began its work in December 2010. The study was to be completed in 2 years. The interim report was provided to DOE on March 2, 2012, and released to the public on March 7. The interim report states that its 4 meetings leading up to the interim report "were concerned mainly with information gathering through presentations, and the committee is only now carrying out the detailed analysis of the many important topics that will be included in its final report."

The interim report states, "Important topics that are not addressed in this interim report—but will be addressed to the extent possible in the final report—include an analysis of the cost-effectiveness of inertial fusion energy, a comparison of the various driver options, and an R&D roadmap at the conceptual level for a national program aimed at the design and construction of an inertial fusion energy demonstration plant, including approximate estimates, where possible, of the funding required at each stage. At the outset of the study, the committee decided that the fusion–fission hybrid was outside the scope of the study."

The interim report provides two "preliminary" conclusions and one recommendation, as follows:

Conclusion 1: The scientific and technological progress in inertial confinement fusion has been substantial during the past decade, particularly in areas pertaining to the achievement and understanding of high-energy-density conditions in the compressed fuel, in numerical simulations of inertial confinement fusion processes, and in exploring several of the critical technologies required for inertial fusion energy applications (e.g., high-repetition-rate lasers and heavy-ion-beam systems, pulsed-power systems, and cryogenic target fabrication techniques).

The report states, "Despite these advances, however, many of the technologies needed for an integrated inertial fusion energy system are still at an early stage of technological maturity. For all approaches to inertial fusion energy examined by the committee (diode-pumped lasers, krypton fluoride lasers, heavy-ion accelerators, pulsed power; indirect drive and direct drive), there remain critical scientific and engineering challenges associated with establishing the technical basis for an inertial fusion energy demonstration plant."

Conclusion 2: It would be premature at the present time to choose a particular driver approach as the preferred option for an inertial fusion energy demonstration plant.

The report states, "The committee recognizes, of course, that such a down-selection among options will eventually have to be made. In its final report, the committee will provide examples of key experimental results that will be needed to inform the decision points regarding which driver-target combinations are most likely to succeed."

The report notes, "The U.S. Department of Energy's (DOE's) National Nuclear Security Administration (NNSA) supports a major national effort in inertial confinement fusion at the National Ignition Facility (NIF) that is focused primarily on addressing technical issues related to stewardship of the nation's nuclear weapons stockpile and national security. An intense national campaign is underway to achieve ignition conditions on the NIF, and there has been considerable initial technical progress toward this major goal, although progress has been slower than originally anticipated."

The report states, "The current NIF laser, targets, shot repetition rate, production methods, and materials are not specifically designed to be suitable for inertial fusion energy (IFE) applications. Nevertheless, many experiments that could be done using the NIF would be valuable for IFE even if the achievement of ignition is delayed—particularly those that provide experimental validation of predictive capabilities."

The report states, "The above discussion led the committee to make the following recommendation."

Recommendation: Planning should begin for making effective use of the National Ignition Facility as one of the major program elements in an assessment of the feasibility of inertial fusion energy.

Dr. Steven Koonin, the DOE Undersecretary for Science, requested the Academies' review of inertial fusion for energy. However, Koonin resigned his post in November 2011, resulting in considerable doubt that there would be a receptive "audience" for the Academies' recommendations at the DOE.

As of September 2012, the Committee's final report had received classification clearance from DOE and was undergoing internal National Academies' review prior to release.

The FY 2013 US Fusion Program Dilemma

President Obama sent his FY 2013 budget request to Congress on February 13, 2012. For the Department of Energy, the President requested $27.155 billion compared to the FY enacted budget of $26.3 billion, a 3.2% increase.

For the DOE Office of Science, the President requested $4.992 billion compared to the FY 2012 enacted budget of $4.874 billion, an increase of 2.4%. For the DOE Office of Fusion Energy Sciences, the President requested $398.3 million compared to the FY 2012 enacted budget of $401 million, a decrease of 0.7%. However, within the total, the US contribution to the international ITER project would increase from $105 million to $1 50 million, resulting in a decrease in the US domestic base fusion program to $248.3 million from the FY 2012 level of $296 million, a decrease of 16%. The primary casualty within the domestic program is a proposed termination of the Alcator C-Mod tokamak program at MIT, although most other elements of the domestic program are also slated for reductions.

The budget document noted that DOE's most recent estimate of the total cost of the US contribution to ITER, made in 2008, was a range between $1.4 billion and $2.2 billion, but says, "It is possible that costs may increase beyond (that range)." The budget document says that the cost estimate will be updated in a review this spring (2012). As of September 2012, however, DOE has not released a new total cost estimate or out-year funding profile for US contributions to ITER.

For the Inertial Confinement Fusion Ignition and High Yield Program within the DOE National Nuclear Security Administration (NNSA), the budget proposes $460 million compared to the FY 2012 enacted budget of $474.8 million, a reduction of 3%.

The proposed increase in US ITER spending at the expense of the US domestic fusion program came as a shock to the US fusion community. ITER was supposed to be a high-level commitment among the governments of the seven ITER Parties. On January 30, 2003, President George W. Bush had issued a statement saying, "I am pleased to announce that the United States will join ITER, an ambitious international research project to harness the promise of fusion energy"'. The results of ITER will advance the effort to produce clean, renewable, and commercially-available fusion energy by the middle of this century." On the same day, in a speech to fusion researchers at the Princeton Plasma Physics Laboratory, US Energy Secretary Spencer Abraham had said, "But let me be clear, our decision to join ITER in no way means a lesser role for the fusion programs we undertake here at home. It is imperative that we maintain and enhance our strong domestic research program—at Princeton, at the universities and at our other labs. Critical science needs to be done in the USA, in parallel with ITER, to strengthen our competitive position in fusion technology."

The US fusion community reacted strongly to the proposed cuts in the domestic fusion program. In a February 27, 2012 letter sent to Energy Secretary Steven Chu and President Obama's Science Advisor John Holdren, seven senior members of the community (David Anderson, Ray Fonck, Stan Milora, Miklos Porkolab, Stewart Prager, Ned Sauthoff, and Tony Taylor) said, "As leaders and stewards of the current U.S. fusion research effort, we are unfortunately compelled to point out that the Fiscal Year 2013 budget request will demote the U.S. program to a second-tier player in the world fusion effort." The letter states, "After years of operating on minimal budgets and essentially level funding, the domestic fusion program cannot withstand the proposed reductions without severe negative impact to our essential capabilities and our scientific contributions to the international fusion program and ITER." "If implemented," the letter says, "the $49 million cut contained in the budget request will result in the layoff of hundreds of fusion scientists, engineers, graduate students, and support personnel…."

The letter states, "The President's budget request, while holding the total fusion effort approximately level ($398 M compared to $401 M in FY 2012), cut the domestic fusion effort by $49 million, while increasing the US contribution to ITER construction by $45 million (to $150 M compared to $105 M in FY 2012). The proposed cuts to the domestic program would affect just about every sub-element of the program: tokamaks, high energy density physics, theory and computation, general plasma physics, plasma technology and advanced design (systems studies)."

An article in the February 24, 2012 issue of Science magazine by Adrian Cho, headlined "Bigger Contribution to ITER Erodes Domestic Fusion Program," summarizes the reaction of the fusion community. "I was shocked," the article quotes MIT's Porkolab as saying, "I didn't have the faintest idea of what was coming." (The budget proposes to shut down the Alcator C-Mod at MIT, one of the three major fusion research tokamaks in the USA.) The article quotes Martin Greenwald, chair of DOE's Fusion Energy Sciences Advisory Committee, as saying that contributing to ITER "is reasonable only in the context of a domestic program. Otherwise you're just building a piece of equipment for other people to use." Princeton's Stewart Prager is quoted as saying, "If all the cuts go through, we would have to lay off about 100 of 435 staff."

At its February 28–29, 2012 meeting, the DOE's Fusion Energy Sciences Advisory Committee (FESAC) also provided comments on the Department's FY 2013 proposed fusion program budget that indicated plans to make significant cuts in the domestic fusion budget. At the meeting, FESAC members expressed dismay that the DOE had not discussed the proposed actions with FESAC or the fusion community.

In a letter to DOE Office of Science Director William Brinkman, FESAC chair Martin Greenwald wrote, "It is clear that the community is upset about the current budget trajectory and the potential impact on our domestic program." Greenwald said, "I felt that it was important for you to understand the views expressed by the committee without delay." He said that the FESAC had voted 17–0, with 2 recusals and 1 absent to tell him that "The committee objects to the theme/impression that these cuts leave the program relatively unscathed and strongly cautions against claims of impactful potential at this level or lower without real study and discussion." The FESAC statement says, "the damage is real" and the "portent for the future is even more threatening." The statement says that if the ITER situation "is in flux inside the Administration, this does not appear to be the time to make termination decisions that cannot be reversed." It questions "the wisdom to do lasting changes to program based on an undefined ITER (funding) profile and in the absence of an overall plan for the program."

The FESAC urged DOE to seek "buy-in" from the fusion community, saying, "cohesion of (the fusion) community is critical as we confront hard decisions." They say, "we don't want (the) community to give a message different from DOE/OS/OFES." The FESAC urged Dr. Brinkman to request assistance from FESAC in addressing "near-term crisis management, shoring up the case for domestic research while ITER is under construction." They state, "We cannot maintain a viable fusion science program on flat $400 M budget," noting "Once a field is shut down, you need to start over and that can take decades." They ask Brinkman to charge FESAC to assist DOE to prepare a fusion plan. They state, "The plan should run to 2021 (ITER start): and include option(s) and plans for next decade." They say the study should include "planning for ITER-era Burning Plasma leadership" and "planning for (a) Fusion Nuclear Science Program leading to fusion energy."

At the February 28–29, 2012 FESAC meeting, I provided a statement in the "public comment" portion of the meeting, which read, in part:

"First, let me say that I endorse the recommendation just made by Dr. Earl Marmar of MIT that no irrevocable decisions be made relative to reductions in the fusion program, as proposed in the President's FY 2013 budget submission to Congress, until a vetting of such reductions occurs within the U.S. fusion community. This should be done by FESAC, or otherwise, to seek community consensus relative to priorities identified previously by FESAC."

"Much of the discussion has been focused on the proposed termination of the Alcator C-Mod program at MIT. The proposed termination is of serious concern, since that program has made, and is making, important contributions to our understanding of tokamak physics and, furthermore, is important to the training of the next generation of fusion scientists. Termination of Alcator C-Mod would mean a "double whammy" for the MIT fusion program, since DOE terminated the other significant experimental facility there last year, the Levitated Dipole Experiment (LDX). Without these two facilities, MIT will lack the facilities to continue providing experience to students doing experimental fusion research."

"But the problem with the proposed reductions is much broader and more serious than just the role and future of the MIT program. Reductions in other areas, such as High Energy Density Laboratory Plasmas (HEDLP), theory, and systems studies will result not only in a loss of valuable talent and expertise throughout the U.S. fusion program, but will also mean that research results these people and facilities would otherwise provide in the coming years will not obtained."

Table 10.3 US ITER project funding profiles

US ITER project funding	Original profile (2007) $M	Current profile (2012) $M
FY 2006	19.3	25.0
FY 2007	60.0	60.0
FY 2008	160.0	10.7
FY2009	214.5	124.0
FY 2010	210.0	135.0
FY 2011	181.3	80.0
FY 2012	130.0	105.0
FY 2013	116.9	150.0
FY 2014	30.0	?
FY 2015	Project complete $1.122 B	?
FY 2016		?
FY 2017		?
FY 2018		?
FY 2019		?
		Project complete>$2 B

? not known

I said, "The reductions proposed in the domestic fusion program were deemed necessary by DOE in order to increase funding for the U.S. contribution to ITER from $105 million in FY 2012 to $150 million in FY 2013. As several FESAC members noted yesterday, we have not been told by DOE how much is really needed in FY 2013, how much will be needed in future years to meet the November 2019 ITER first plasma target date, or where these funds will come from. We were told yesterday that Japan plans to spend $250 million in 2013 to maintain their ITER commitment. Since the US has the same one-ninth share of ITER commitments, it would seem logical that the US may really have needed roughly that amount in order to meet the ITER schedule. Thus, even with $150 million in FY 2013, the US may not have the funds it really needs for ITER in FY 2013."

Table 10.3 shows the funding profiles required for the US ITER project contribution, as provided by DOE to Congress in early 2006 in the FY 2007 budget request, and the actual funds provided and requested as of 2012. DOE had not provided updated profiles as of summer 2012 to meet the scheduled November 2019 project completion date.

I said, "Thus, if US ITER receives the requested $150 M in FY 2013, it will have spent $690 M through FY 2013. If ITER is to operate in November 2019, essentially all needed construction funds must be spent by end of FY 2018. Since the latest (informal) estimate of the total US contribution to ITER has risen reportedly to $2.6 billion, the President will need to request, and the Congress will need to appropriate, an additional nearly $2 billion over the five fiscal years 2014–2018, or an average of nearly $400 million per year. Clearly this cannot come by continuing to decrease the US domestic fusion program. Something needs to be done."

I noted, "On January 30, 2003, the U.S. decided to rejoin the ITER project. The decision was made at the highest level of the US government, an announcement from

President George W. Bush stating, 'I am pleased to announce that the United States will join ITER, an ambitious international research project to harness the promise of fusion energy.' " I said, "To ensure the successful completion of the ITER project, without destroying the U.S. domestic program, requires that we regain the high level U.S. government support for the project that seems to have been lost in the FY 2013 budget submission. The ITER project must be again recognized as a presidential commitment that cannot be funded by reducing the U.S. domestic fusion effort."

I concluded, "The U.S. domestic fusion program does not have sufficient funds to pay for the U.S. contribution to ITER construction."

ANS Fusion Energy Division Comments on FY 2013 Budget Proposal

On April 10, 2012, a statement prepared by the American Nuclear Society (ANS) Fusion Energy Division on the President's FY 2013 fusion budget was sent to key members of the House and Senate Appropriations Committees, with copies to senior officials in the executive branch. The statement says, in part:

> The FY-2013 budget request by the Administration endangers the United States' domestic fusion program as well as our country's scientific contributions to the ITER international project. If implemented, the FY 2013 budget reductions will deal a major blow to the U.S. fusion research program and further erode its leadership position. After years of operating on minimal budgets and essentially level funding, the U.S. fusion program cannot withstand the proposed reductions without significant negative impacts.
>
> U.S. fusion researchers were told a few years ago that there would be some "belt tightening" to divert fusion research funds to ITER construction. Without any quantitative guidance from the DOE on belt-tightening, there was speculation that it might be 1% or perhaps even as much as a 5% budget reduction for a few years. The FY 2013 budget, however, proposes a 16% reduction ($45 million) of fusion research funds, and DOE officials have given warnings that reductions of up to $100 million more will be needed in the coming years. If the Administration's FY 2013 budget is implemented, the DOE will close a unique fusion experiment, the Alcator machine at MIT, and the students and staff there will be dispersed. Deeper cuts in the future will disperse even more staff and students at institutions around the country who would use the ITER results, and greatly reduce the number of American engineers and scientists who will be educated and trained in fusion.
>
> We urge the U.S. to consistently and adequately support the fusion research program as outlined in the Energy Policy Act of 2005 (PL 109–58, sec 971–972) and reverse this position, restoring funds to the domestic fusion program budget and, separately, fully funding this nation's promised annual ITER contribution.
>
> The path to discover commercially viable fusion energy is one of the grand scientific challenges of our time. With ITER under construction to explore the science of burning plasmas, the world fusion program is poised to enter its final era of research. Other nations, including China, the European Union, Japan, Russia, and South Korea, are forging ahead rapidly, investing heavily in their domestic fusion programs and in educating the next generation of fusion researchers. They are fully supporting ITER as well. The U.S. has consistently led the fusion field and should continue to do so. American leadership in fusion energy would be in the best interests of the U.S. and science itself.

FESAC Launches Another Priorities Study

The US magnetic fusion community was clearly unhappy with the domestic fusion program cuts proposed in FY 2013 budget sent to Congress by the DOE. They asked that the community be consulted on the distribution among subprograms of any eventual reductions. Consequently, the DOE Fusion Energy Sciences Advisory Committee (FESAC) was asked to provide a new assessment "of priorities among and within the elements of the magnetic fusion energy sciences program." The study was requested by DOE Office of Science Director William F. Brinkman in the wake of President Obama's Fiscal Year 2013 budget proposal to reduce funding for the (non-ITER) US domestic fusion effort by $48 M (16%). However, Brinkman stated in an April 13, 2012 charge letter to FESAC, "Note that General Plasma Science and High-Energy-Density Laboratory Physics programs and US contributions to ITER are not part of this charge." Thus, DOE's charge to FESAC fell far short of FESAC's request to review the entirety of the fusion program funding priorities for FY 2013 and beyond.

FESAC established a subpanel under the chairmanship of FESAC member Bob Rosner (U, Chicago). They were asked to "prioritize among and within the FY 2013 elements of the non-ITER magnetic fusion portion of the Fusion Energy Sciences program" under three budget cases: (1) "the 2013 Presidential budget request level of effort," (2) "a restoration of the budget to the 2012 level for that part of the program," and (3) "a program that has a substantially enhanced emphasis on fusion materials science."

The subpanel received presentations at several public meetings and also received "white papers" from the fusion community. These, along with the charge letter and list of subpanel members, are posted at http://fire.pppl.gov.

Brinkman gave the FESAC no target date to complete the study, but FESAC hoped to complete the study and provide its recommendations by early fall 2012.

Congressional Actions

In April 2012, Appropriation Committees in the US House of Representatives and the US Senate marked up President Obama's FY 2013 budget request for the Department of Energy and sent starkly different messages to the US fusion program.

The Senate endorsed the President's proposed fusion budget, including drastic cuts in the domestic fusion effort, transferring that money to growth in spending for the US contribution to ITER construction, all within an overall fusion budget that is only slightly below that of FY 2012. The House, on the other hand, flatly rejected the proposed cuts to the domestic fusion budget and, in addition, provided more funds for ITER than the President requested. The Senate bill provides $398 million for the Office of Fusion Energy Sciences, whereas the House bill provides almost $475 million.

The report language accompanying the House and Senate markups for both the Office of Fusion Energy Sciences (within DOE's Office of Science) and the Inertial Fusion Ignition and High Yield Campaign (within DOE's National Nuclear Security Administration weapons budget) is given below.

With respect to OFES, the House said:

The Fusion Energy Sciences program supports basic research and experimentation aiming to harness nuclear fusion for energy production. The Committee recommends $474,617,000 for fusion energy sciences, $72,440,000 above fiscal year 2012 and $76,293,000 above the request.

The domestic fusion program is a critical component of United States science leadership and a necessary building block of any successful fusion projects, including ITER. The recommendation includes $296,617,000 for the domestic fusion program, $560,000 below fiscal year 2012 and $48,293,000 above the request. The request proposes to shut down the Alcator C-Mod facility and provides only enough funding for decommissioning and existing graduate students. The Department is instead directed to continue operations at the Alcator C-Mod facility and to fund continued research, operations, and upgrades across the Office of Science's domestic fusion enterprise.

The recommendation includes S178,000,000 for the United States contribution to ITER, the international collaboration to construct the world's first self-sustaining experimental fusion reactor, $73,000,000 above fiscal year 2012 and $28,000,000 above the request. ITER is an important international collaboration that represents a major step forward in fusion energy science, but its funding requirements will create substantial budgetary challenges throughout the decade. The Committee appreciates that the Office of Science is grappling with these challenges but notes that the budget request does not propose a viable or well-planned solution. The Committee recommendation includes funding to continue the domestic fusion program at approximately the fiscal year 2012 level, and to increase ITER towards its planned funding level for fiscal year 2013. Looking forward, however, the increasing requirements for ITER will continue to pose challenges, and the Committee believes that long-term policy decisions for the Fusion Energy Sciences should be guided by impartial analysis of scientific needs and opportunities, and with an eye on American competitiveness and leadership. The Committee therefore reiterates the importance of the 10-year plan for Fusion Energy Sciences directed in the fiscal year 2012 appropriations conference report, of that plan's timely delivery, and of the inclusion of priorities across domestic and international fusion facilities, projects, and programs.

With respect to NNSA's Inertial Confinement Fusion program, the House said:

The Committee recommends $480,000,000, $3,726,000 above fiscal year 2012 and $20,000,000 above the budget request. Within these funds, $62,500,000 shall be for the OMEGA Laser Facility at the University of Rochester, $2,250,000 above the request.

As the first ignition campaign comes to a close in fiscal year 2012, it is a distinct possibility that the NNSA will not achieve ignition during these initial experiments. While achieving ignition was never scientifically assured, the considerable costs will not have been warranted if the only role the National Ignition Facility (NIF) serves is that of an expensive platform for routine high energy density physics experiments. The Committee continues to support the pursuit of ignition and urges the NNSA to develop a cost-effective strategy for future experimental activity as the next phase of scientific effort begins. The recommendation supports a lower, though still robust, level of experimental activity on the NIF in fiscal year 2013 given the completion of major diagnostic acquisitions and the shift in experimental tempo.

Further, the Committee supports the application of a fair and standardized overhead rate that fully adheres to proper cost accounting standards. In previous years, the NNSA allowed Lawrence Livermore National Laboratory to apply a reduced overhead rate for the operation of the NIF which artificially lowered the amount of funding needed within the ICF Campaign to conduct experimental activities, in violation of cost accounting standards.

This practice misrepresented the full costs of these activities and shifted those costs onto other programs at the laboratory. While the ultimate programmatic impacts of the rate shift are still not clear, there is flexibility within the NNSA budget to partially mitigate those consequences as the overhead rate transitions back to a more appropriate level. Nevertheless, it is apparent that the NNSA did not properly take into account those impacts when developing its budget request and the Committee recommends $20,000,000 above the request to mitigate any unintended adverse impacts in fiscal year 2013. The Committee will continue to work with the NNSA to understand the implications of the transition to an appropriate overhead rate at the NIF and adjust resources as necessary so the facility may effectively execute its mission.

With respect to OFES, the Senate said:

The Committee recommends $398,324,000 as requested for Fusion Energy Sciences. Within these funds, the Committee recommends $150,000,000 as requested for the U.S. contribution to ITER.

Similar to the Nuclear Physics program, the Committee is concerned by the lack of strategic direction for the fusion energy program. The Committee understands that the budget request provides a $45,000,000 increase to the U.S. ITER contribution but even with the increase, the U.S. contribution is still $50,000,000 short of the project plan.

The Committee also understands that the increase to the U.S. contribution came at the expense of the domestic fusion program. The Committee is concerned that additional cuts to the domestic fusion energy program may undermine U.S. advances in fusion and the U.S. ability to take advantage of scientific developments of the ITER project. The Office of Science believes that it can take advantage of international programs and facilities to build and maintain U.S. expertise in fusion energy sciences. However, a February 2012 Fusion Energy Sciences Advisory Committee report cautioned that international facilities in Asia and Europe will not be operating for several more years and international "collaborations cannot come at the expense of a domestic research program that can benefit from ITER.

The Committee directs the Office of Science to assess the impact to the domestic fusion energy sciences workforce and the ability of the United States to take advantage of ITER to advance fusion energy before recommending any further cuts to the domestic program. The Committee also directs the Office of Science to assess alternatives to participating in the ITER project, including reducing contributions to the project, and the impact of withdrawing from the project, if necessary, to maintain domestic capabilities. Further, the Committee directs the Office of Science to include a project data sheet with details of all project costs until the completion of the project for ITER in the fiscal year 2014 budget submission.

The Committee understands that DOE provides funding for ITER as a Major Item of Equipment rather than a line item construction project, which would be consistent with DOE Order 413.3B. However, the Committee feels that a multi-billion dollar project, especially of this scale and complexity, should be treated as a construction project and follow DOE Order 413.3B guidance.

With respect to NNSA's Inertial Confinement Fusion program, the Senate said:

The Committee recommends $460,000,000 as requested. The Committee understands the importance of the National Ignition Facility [NIF] and supports NNSA's efforts to ensure the long term viability of the facility when the National Ignition Campaign ends. The Committee encourages NNSA to work closely with the Lawrence Livermore National Laboratory to help manage the required full transition of the facility to the laboratory's standard cost accounting practices. The Committee directs NNSA, with congressional

> notification to the House and Senate Appropriations Committees, to use up to $140,000,000 of Lawrence Livermore National Laboratory's internal additional direct purchasing power—generated by the overall lowering of the laboratory's "Blended Rate" resulting from NIF's transition away from a Self Constructed Asset Pool indirect rate and reduced management fee—to increase the level of the laboratory's Readiness in Technical Base and Facilities funds dedicated to supporting NIF. The Committee recommends that NNSA move the NIF operating budget to the Readiness in Technical Base and Facilities budget line, which would be consistent with the facility's transition to regular operations and how other facilities are funded. The Committee also recommends that NNSA consider alternatives to operating the facility 24 h a day, 7 days a week.
>
> Also within the funds for inertial confinement fusion, at least $62,000,000 and $55,000,000 shall be used for inertial confinement fusion activities at the University of Rochester's Omega facility and Sandia National Laboratory's Z facility, respectively. The Committee also recommends at least $5,000,000 as requested for the Naval Research Laboratory to continue operating laser facilities focused on laser plasma interactions, target hydrodynamics, materials, and advanced ignition concepts.
>
> The Committee remains concerned about NIF's ability to achieve ignition—the primary purpose of constructing the facility—by the end of fiscal year 2012 when the National Ignition Campaign ends and the facility is to transition to regular ignition operations and pursue broad scientific applications. The Committee directs NNSA to establish an independent advisory committee as soon as possible to help set a strategic direction for inertial confinement fusion and high-energy density physics research and determine how best to use current facilities to advance this scientific field. If NIF does not achieve ignition by the end of fiscal year 2012 using a cryogenically layered deuterium and tritium target that produces a neutron yield with a gain greater than 1, the Committee directs NNSA to submit a report by November 30, 2012 that (1) explains the scientific and technical barriers to achieving ignition; (2) the steps NNSA will take to achieve ignition with a revised schedule; and (3) the impact on the stockpile stewardship program.
>
> To meet the complex and increased mission requirements of the Inertial Confinement Fusion and Science Campaigns at a period of constrained funding, the Committee urges the Department to continue its activities to ensure a multiple vendor base capable of cost-effectively developing and fabricating the full range of targets for inertial confinement fusion facilities that support the stockpile stewardship program.

The House Appropriations bill subsequently was passed by the full House. Before becoming law, however, the Senate must vote to approve an Appropriation bill; the bills then must go to a House–Senate conference committee to resolve differences, and then both House and Senate must pass the same bill and send it to the President for signature. As of mid-September, the Senate had taken no vote on the bill and it seems unlikely an Appropriations bill will reach the President's desk until after the November 2012 elections. Fiscal years begin in October 1 of each year.

ITER Council Meets in Washington

The ITER Council met in Washington, DC, on June 20–21, 2012. The Council acknowledged a number of "positive advancements for the project, noting, in particular, the progress of ITER construction and licensing." The Council stressed that "respecting the project schedule within cost remains a critical issue and that reported slippages need rapid correction; it also noted that the ITER Organization and the seven Domestic Agencies are working together on these corrections." The Council "expressed concern over delays in the manufacturing of some major components such as the vacuum vessel." Council members "reasserted the necessity of keeping the project well on schedule and within cost."

Uncertainties

As summer 2012 drew to a close, uncertainty reigned on several fronts. A presidential election campaign was under way, with the resulting uncertainty of whether President Obama would be reelected in November. Congress had not yet been passed the FY 2013 budget. The NIF had not yet ignited a fusion capsule. The National Academies' inertial fusion energy review panel had not yet issued its final report. The FESAC review of magnetic fusion funding priorities had not yet been completed. The DOE had still not released its estimate of the ITER US future year funding requirements. The world economy was still facing an uncertain future. And the US Congress was in gridlock over how to foster US economic growth while making deep cuts in federal spending and also while simultaneously continuing growth in defense spending.

Applications 11

The best way to predict your future is to create it.

Peter Drucker
Author, Managing the Future

Electric Power

The primary (and perhaps the most difficult) goal of the fusion program has been the production of base-load electricity. Like nuclear fission power plants, fusion power plants are unlikely to lend themselves well to situations that require rapid changes in power levels. This characteristic, as well as economies of scale, tends to make fusion reactors more likely to be economically competitive at relative large sizes in central station (as distinguished from distributed) electricity generation. The latter (distributed) future market most likely will be filled by an array of solar, wind, and possibly gas turbine technologies. The primary competition for fusion will, then, be from coal, nuclear, and possibly natural gas power plants.

Electric vehicles are a small but growing segment of the transportation system. If electric vehicles replaced all gasoline-powered vehicles, the electric generating capacity of the country would have to be approximately doubled. Such a situation would radically change the need for both nuclear fission and fusion power.

Until the mid-1980s, US electric utilities were highly regulated by local and state government utility boards that approved both rates and what costs could be included in the rate base. The boards typically encouraged and approved a small charge for research and development. Consequently, most large utility companies had research departments, and individuals in these departments often developed close relations with nearby fusion research laboratories. In some cases, individual utilities, like Pacific Electric and Gas in the San Francisco Bay Area, funded small independent fusion research efforts. Also, the utilities sponsored independent fusion research efforts via the Electric Power Research Institute. All of this changed in the mid-1980s when utilities were "deregulated" and told to compete among themselves for

S.O. Dean, *Search for the Ultimate Energy Source: A History of the U.S. Fusion Energy Program*, Green Energy and Technology, DOI 10.1007/978-1-4614-6037-4_11,

customers. The result was that utility executives embarked on cost cutting in order to offer reduced rates to potential customers. The research departments at individual utilities were a major casualty of this change in policy. The EPRI Board of Directors also decided that EPRI should focus on research aimed at lowering the cost of electricity of its member companies in the near term rather than conducting research aimed at developing new long-term energy sources like fusion. EPRI abandoned its small (about $4 million per year) fusion effort.

Hydrogen Production

One hears much these days about the future potential of using hydrogen as direct fuel for commercial energy or, more likely, in combination with fuel cells for transportation. Today, hydrogen is obtained by a chemical process from hydrocarbons (primarily natural gas), which leaves an environmental issue of what to do with the residual carbon compounds. The technology for producing hydrogen is already in widespread use in the petrochemical industry.

Hydrogen can also be obtained from water, though less efficiently than from hydrocarbons, by electrolysis. As the name electrolysis implies, this requires a large source of electric power as part of the process. If a large commercial market were to develop for hydrogen, fusion would be a logical competitor with other power sources, to provide the electricity required for electrolysis. The duty factor required of a fusion plant could be relaxed compared to central station power requirements since the hydrogen can be stored. This would allow operators to gain experience with early fusion plants that might have low duty factor. The advantage of fusion (and fission) compared to natural gas for large-scale hydrogen production comes from the environmental benefit of not producing carbon emissions. The advantage relative to using solar or wind comes from being more capable of producing larger quantities of hydrogen in larger plants.

Hydrogen production and its potential use in transportation in connection with fuel cells are discussed in some detail by Chen [107]. A FESAC study [108] looked at hydrogen production and many of the other nonelectric applications of fusion discussed below.

Fusion–Fission Hybrid Reactors

The idea of a fusion–fission hybrid reactor is to surround the fusion core plasma region with either a blanket of subcritical fissionable material or a blanket of nonfissionable uranium-238 or thorium-232. Several applications can be envisaged depending on design choices. A low-performance fusion core could drive a subcritical, therefore "safer," fission blanket to produce electricity, or neutrons from the low-performance fusion core could be used to produce fuel for the current generation of light water fission reactors as described in the next section. The design could also be optimized to convert nuclear fission reactor waste into shorter-lived or nonradioactive elements. A DOE-sponsored review [109], chaired by Jeffrey Freidberg, concluded, "the fusion–fission hybrid does indeed offer the promise of being able to address the sustainability issues associated with conventional nuclear power." But, they said, "on the other hand, when participants (in the study) were asked whether these hybrid solutions are potentially more attractive than contemplated pure fission solutions (that is, fast burners and fast breeders), there was general consensus that this question could not be quantitatively answered based on the known technical information."

A review of fusion–fission hybrid reactors was recently published in the Wiley Nuclear Energy Encyclopedia [110]. A study based on a laser inertial confinement fusion hybrid called LIFE has been underway at the Lawrence Livermore National Laboratory for the past few years [111]. Wallace Manheimer has also been an articulate advocate of the fusion–fission hybrid [112].

Fuel for Nuclear Fission Reactors

Several studies [113, 114] have shown that neutrons from a single fusion reactor could be used to produce enough fissionable material (e.g., uranium or plutonium) from nonfissionable material to fuel 5 or more light water fission reactors. The advantage of this application from the fusion viewpoint is that the fusion reactor would have less demanding duty cycle requirements than needed for an electric power plant. Currently, there is an adequate supply of fissionable material at reasonable cost to fulfill the fuel requirements of today's fission power plants. This situation would have to change radically (e.g., by a large increase in fission reactor deployment or a large increase in the cost of uranium) before there would likely be a market for this application of fusion. Also, fission reactor designs have evolved in such a way as to make it less likely that fusion would be needed for this application any time soon.

Conversion of Nuclear Waste

Today's nuclear fission power plants produce large amounts of radioactive waste. This material consists of a variety of hazardous materials, including short- and long-lived fission products and unfissioned fuel. In some countries, this waste is reprocessed so that the unfissioned fuel is recaptured and can be reused, thus reducing the amount of fission products requiring long-term storage or disposal. In the United States, no reprocessing is in use and the waste is being stored on-site at existing power plant sites. Plans to provide for deep geological storage facilities are on hold. What to do about fission power plant waste is a major impediment to more widespread use of nuclear fission power, especially in the United States.

Use of fusion or accelerator-produced neutrons to convert long-lived fission products into either shorter-lived or stable isotopes is one of several technologies that have been studied for this purpose [115–117]. As is the case for the production of fuel for fission reactors, one of the advantages from the fusion viewpoint is that the fusion reactor would not have the duty cycle demands required for electricity production. Another advantage is that the performance requirements on fusion energy production from the plasma are lessened and comparable to the performance in today's fusion experiments.

Processing of Other Wastes

In early 1969, US Atomic Energy Commission physicists William C. Gough and Bernard J. Eastlund issued the first in a series of reports [118]: "The Fusion Torch: Closing the Cycle from Use to Reuse." They describe how plasmas, in general, and fusion plasmas, in particular, could be used to break down materials currently piling up in waste dumps into reusable materials and even possibly back into their elemental form.

The idea of the fusion torch is to use ultrahigh temperature plasmas to vaporize, disassociate, and ionize solid wastes in such a manner as to permit the recovery of the elements and compounds for reuse. They note that many of the ancillary technologies, such as electromagnetic separators and centrifuges, that might be incorporated into a working system are already in use for other purposes.

Plasma torches are already in commercial use for processing waste; fusion torches must await the availability of a fusion source.

Desalination

Sources of freshwater are in short supply in many regions of the United States and even more so in other parts of the world. This problem will grow more acute as world population continues to grow. Fortunately, there is a huge supply of salt water in the oceans which can be desalinated easily (though expensively) by evaporation and condensation (distillation) and other methods. To produce large quantities of freshwater this way does require large amounts of heat such as can be produced by the same technologies (coal, fission, fusion, etc.) that can be used for large central station power plants. The technology to do this is available, though in most cases it is not economic if adequate natural sources of freshwater are available. Nuclear aircraft carriers, for example, use the heat from their nuclear power plants to desalinate water, and large-scale desalination plants are operational in the Arab Emirates and elsewhere using fossil fuel plants as the heat source. In the future, it is to be expected that the economics for desalination as a source of freshwater will become more and more attractive and even, in some cases, absolutely necessary.

For fusion, the possibility of a desalination market has, once more, the advantage that "duty factor" will not be as critical as it is for electricity generation, where the supply of electricity must be continuous. For water, large quantities can be stockpiled to account for planned or unplanned outages of the desalination plant.

Studies and actual experience in designing and operating nuclear- and fossil-based desalination plants are largely applicable to a fusion-based plant. A number of studies have been performed specifically analyzing the use of fusion for desalination [119, 120]. From the fusion viewpoint, the advantage of fusion is that fusion produces no greenhouse gases (compared to fossil-based plants) and may have better economics (compared to fission) at very large scale.

Spin-Offs

Although fusion for the production of electricity or other applications remains in the future, much of the science and technology developed in conjunction with fusion R&D is already having great commercial impact. These applications include such commercial markets as plasma etching of semiconductor-integrated circuits; ion implantation for hardening of tools, dies, and industrial metals; printing of polymer films; deposition of anticorrosion and other types of coatings; production of high-performance ceramics; treatment of hazardous waste; surface cleaning of materials; laser and pulsed X-ray applications; plasma flat-panel displays; improvements in materials for a variety of applications; high-current switches for the power industry; medical and biological applications; improvements in a wide variety of related technologies, including isotope separation, microwave sources, cryogenics, superconductivity, and optics; new technologies such as light sources and digital radar; and contributions to many areas of basic science such as space physics and supercomputing. The current world market for these applications is conservatively estimated to be in excess $200 billion per year [121–124].

12 Engineering Challenges

Only when Edison realized that his lightbulb would have to work outside the lab and be part of a total system did it dawn on him how to create a properly working bulb.

–Ira Flatow
Author, They all Laughed… From Lightbulbs to Lasers

The focus of the US fusion program throughout its history has been the achievement of what has often been termed "scientific feasibility," i.e., producing and containing a high temperature plasma that exceeds the Lawson criterion. Facilities like the Tokamak Fusion Test Reactor (TFTR) and the Joint European Torus (JET) have come close to demonstrating that goal, but such achievements, while necessary, are not sufficient to demonstrate the capability to actually produce commercial electric power or any of the other possible commercial applications of fusion discussed in the last chapter. It is thus necessary to couple the production of a fusion plasma to the engineering systems necessary to reliably and cost-effectively transform the fusion energy into a usable form.

The technologies required to do this are legion, and all require a serious development effort. Though the requirements are well understood, the necessary development programs have largely taken a backseat to the needs of the fusion plasma physics effort.

Materials

The fusion plasma is created in a vacuum, but this vacuum exists inside a material chamber. The inside wall of this chamber is often called the "first wall," and other components inside the chamber in proximity to the first wall are usually called "plasma-facing components (PFCs)." The first wall and the plasma-facing components are subjected to a steady stream of neutrons, plasma particles, and other types of radiations from the fusion plasma, and depending on the properties of the materials

S.O. Dean, *Search for the Ultimate Energy Source: A History of the U.S. Fusion Energy Program*, Green Energy and Technology, DOI 10.1007/978-1-4614-6037-4_12,

chosen for the wall and PFCs, they present the first engineering challenge for practical fusion power. Material will be eroded from the surface of these materials and, if that material enters the fusion plasma, will cool it, shutting down the fusion reaction. Furthermore, the material will be damaged by both charged particles and high-energy neutrons from the fusion plasma and will almost certainly have to be replaced periodically over the lifetime of the plant, creating maintenance issues and also adversely affecting the cost of electricity.

Beyond the first wall of the chamber is an area called the "blanket." The function of the blanket region is to slow down the fusion neutrons to capture their energy and convert that energy into heat to drive the steam generators for producing electricity. Another function of the blanket is to capture fusion neutrons in lithium to produce tritium for fuel. This is a demanding job and requires careful choice and arrangement of materials in the blanket. For a superconducting magnetic fusion reactor, a "shield" region is then required between the blanket and the magnets to keep excessive heat from reaching the magnets.

Depending on the choice of materials for the first wall, PFCs, blanket, and shield, neutron absorption results in the production of various types and amounts of radioactive elements. Some materials are deemed more "attractive" than others, though none under consideration are as dangerous, long-lived, or troublesome as those produced in fission reactors. Still, when the useful life of these materials is reached, they must be disposed of as radioactive waste. Considerable effort has been devoted within fusion programs worldwide to identify candidate "low-activation materials" though relatively little effort has been devoted to qualifying them for use in construction. This is partly due to lack of appropriate test facilities and partly due to lower priority assigned to fusion nuclear technology development relative to fusion plasma science. This does not create an issue for today's experimental facilities but will be an issue for fusion power plants. Although the materials needing disposal upon decommissioning of a fusion power plant are shorter lived and less hazardous than from a comparable fission power plant, the volume of such waste from a large tokamak power plant might be comparable, or perhaps even larger, depending on the choice of materials used in construction of the fusion plant.

As fusion reactions continue in the core of a fusion reactor, plasma must continually be removed as exhaust and replenished. The removal in a tokamak is done via a magnetic "divertor." The divertor is considered a PFC, and the material properties of the divertor are, if anything, more demanding than for the first wall and other PFCs.

It is generally conceded that present-day materials widely used in construction are not adequate for use in a fusion power plant. Some steels, which are close to being acceptable, will need to be tested and qualified in facilities that produce 14 MeV neutrons similar in energy, flux, and fluence to be encountered in a fusion power plant. Several types of such facilities have been proposed, but none have yet been built. Some tests in smaller facilities and in fission reactors have been conducted and have provided guidance. It may be possible to avoid or alleviate some solid materials issues. For example, one technique being evaluated in both magnetic and inertial confinement fusion is to employ flowing liquid metals to protect solid walls.

A detailed discussion of fusion materials issues is beyond the scope of this book, but there is extensive literature on this subject [125–128].

Tritium

As discussed in Chap. 1, the easiest fusion reaction to achieve is the fusion of the two heavy isotopes of hydrogen: deuterium and tritium. Deuterium is easily obtained from water and is essentially unlimited in supply. Unfortunately, tritium does not occur in nature in any appreciable amount. This is because it is unstable and decays radioactively with a half-life of 12.3 years. Fortunately, however, it is easily produced from lithium by the capture of a neutron, and lithium is an abundant element, both on land and in the sea [107].

Modest stockpiles of tritium exist, having been produced for nuclear weapons and as a product from heavy water nuclear fission reactors, such as those used for civilian nuclear power plants in Canada and elsewhere. These supplies are, of course, constantly decaying but should be large enough to supply the needs of fusion test facilities such as JET and ITER. They are not, however, large enough to supply the needs of a continuously operating fusion power plant. For this purpose, tritium must be produced in the blanket of the fusion power plant, purified, and fed into the deuterium–tritium fuel mixture of the fusion plasma. The technology for doing this has been analyzed and tested on a small scale. However, it does put additional demands on the choice and arrangement of the materials in the blanket region and adds to the complexity and capital cost of the power plant. In addition to producing tritium, extracting it from the blanket, and arranging for its injection, tritium (along with deuterium and helium) is in the exhaust mixture that continuously leaves the reacting region via the divertor or otherwise. The tritium in this exhaust mixture is very valuable and hence must be separated from the other exhaust gases and recycled back into the fusion plasma. The technology for doing this has been partially demonstrated on a small scale, but engineering challenges for a complete tritium production, extraction, and recycling system still remain. Many of these complexities have been described by Anderson [129] and elsewhere [107] in the literature.

Tritium, being mildly radioactive (it emits a low energy electron that cannot even penetrate the dead layer of the skin), must nevertheless be kept from leaking from the system for safety reasons as well as for its value as a fuel. Tritium diffuses through most materials, so some will become bound in materials that require disposal upon decommissioning of the plant. Engineering the tritium handling system for a fusion power plant will be a major part of the systems design and contribute to plant costs. Engineering of the tritium-related components of a fusion power plant will be challenging but should be manageable [130–133].

Complexity

Fusion power plants will undoubtedly be more complex engineering systems than today's power plants. Complexity will add to both the capital and operational costs of the plant. Hence, if and when more engineering is brought to bear on fusion energy development, significant attention will need to be paid to simplifying the overall system.

Many modern technologies, such as aircraft, are exceedingly complex systems. Some degree of complexity is unavoidable. If the systems are nevertheless reliable and the cost of the complexity is justified by the benefits of the application, complexity is not necessarily unacceptable. Nevertheless, simplification is always to be sought during design and development.

Issues associated with complexity do (or should) affect management decisions on the choice of fusion concepts to emphasize or encourage during the research and development phase. Magnetic confinement configurations come in many variations. Some configurations with more promise based on physics performance seem to require more complex engineering features, whereas some of the less complex geometries seem to have less attractive or less well-understood physics. To a lesser extent, these complexity variations among concepts also affect choices that will have to be made among driver candidates if and when the inertial fusion program begins to construct repetitively pulsed fusion facilities.

Maintenance

Because fusion neutrons will produce radioactive elements in the structural materials of a fusion power plant and because radioactive tritium is used as a fuel, some of which will become absorbed in the structural material, maintenance on much of the fusion vessel will have to be done using remote handling equipment. This requirement will lead to increased complexity in construction and operations and may also require the plant to be off-line for longer time periods compared to today's power plants. All these factors will adversely affect the cost of electricity and affect the ability of fusion power plants to compete with other sources of energy or power.

Remote maintenance is currently being performed in the JET tokamak experiment in the UK, and the technology will be substantially advanced and proven in ITER. Remote handling technologies on a much smaller scale are in use in other industries, so an industry does exist to develop the technologies that fusion power plants will require. Remote handling technologies required for inertial fusion power plants may be simpler than for magnetic plants due to separation of the "drivers" from the fusion chamber.

Cost

The cost of the deuterium and tritium fuel for a fusion power plant is expected to be very low compared to fuel costs in fossil and nuclear fission power plants. However, this advantage will be largely, if not completely, offset by the expected high capital cost of building a fusion power plant, relative to these competing technologies and possibly more costly maintenance costs. Consequently, the actual cost of electricity is predicted in current studies to be comparable. Whether this turns out to be the case depends on many future marketplace factors that cannot be forecast accurately. Among these factors are possible carbon taxes, penalties, or restrictions on the use of fossil fuels; safety, proliferation concerns, or other restrictions on the use of nuclear fission; and the suitability of solar or wind for reliable or large-scale applications.

The capacity factor (or sometimes called the availability) has a large influence on the cost of electricity (COE) from any system. It is a crucial number in converting capital costs to COE. Fusion power systems will be very capital-intensive systems. Such plants will likely operate as base-load power plants where a premium is placed on operating at the maximum capacity factor. Most fusion power plant studies assign a value of typically 70% to 80% to capacity factor. These values cannot be achieved today given the early stages of fusion technology development, so really they represent a goal. By way of comparison, the current fleet of fission power plants in the USA routinely achieves an average capacity factor of about 90%.

Achieving high-capacity factors requires two basic features of the system: high component reliability and acceptable maintenance or downtimes. There is a strong, inverse relationship between the allowed values of the mean time to failure and the mean time to repair for a given component. In other words, the harder it will be to replace the component, the higher must be the degree of reliability. Defining the acceptable values for the mean time to failure and mean time to repair for all the components in a complex fusion power plant will require a comprehensive systems engineering approach.

Achieving high levels of component reliability requires substantial testing and qualification of fusion components, far beyond what has been done to date. For example, no fusion reaction chamber has ever been built, and certainly, none tested to the extent needed to establish failure modes and a reliability database. Given the large number of components and systems in a fusion power plant, a substantial investment in time and money will be required.

The time required to do this testing will have a major impact on the overall timescale to develop commercial fusion systems. At some time, testing in an actual fusion environment will be needed, although much useful testing can and will be done in simulation facilities. Achieving fusion conditions for testing requires very large investments with long timescales and will thus have a profound impact on the roadmap for realizing fusion power systems.

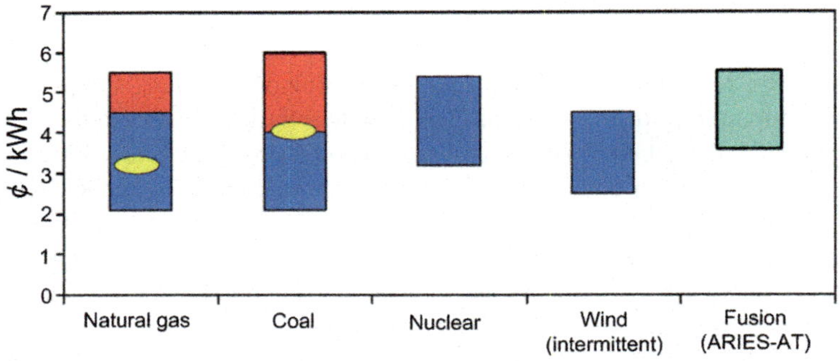

Fig. 12.1 Estimated year 2020 ranges of cost of electricity in the USA in cents per kilowatt hour from different power sources. Yellow ellipses are from US Energy Information Agency and red range is if a $100 per ton carbon emissions tax is imposed on fossil sources [107]

Achieving the necessary replacement times for a fusion system's components is an equally challenging task. Some of these components will require using remote handling systems for maintenance. While the technology and experience in other fields (e.g., fission reactors and space systems) can be adapted to fusion needs, there exists today very limited experience with remote maintenance in actual fusion systems. ITER will be one very important source of such information. Developing the maintenance systems for a fusion power plant will be a significant effort. Unfortunately, there is very little work underway today in the USA on this topic.

For the above reasons, capacity factor probably represents the greatest uncertainty among all the factors that affect the projected cost of electricity for a fusion power plant. This applies to all fusion concepts, both IFE and MFE.

It is difficult to assess capital costs for future fusion power plants since none have yet been built. Therefore, one must be aware that there is a large uncertainty in current projections of both capital cost and cost of electricity for future fusion power plants. Estimates that have been done, however, suggest that fusion could be competitive (Fig. 12.1).

Ultimately, cost of electricity, or cost of other commercial applications of fusion energy sources, will determine whether and when fusion enters the commercial market. The market forces are and will be different in different countries. The US market will likely be the most difficult one in which to compete, as it is more determined by profit goals of the private sector and less by government policy.

Energy 13

There are no diplomatic or military problems associated with energy.
You simply establish a rapid deployment force.
That may cost you a little more than developing energy sources internally but it's a lot more fun.

Edwin E. Kintner
Aide to Admiral Hyman Rickover
and later Head of the US fusion program

Primary Energy Consumption

Energy is measured in a variety of units, most commonly by the unit "BTU" or British thermal unit. One BTU is the amount of energy required to raise the temperature of one pound of water by 1°F. In a home, for example, a furnace might have an output of 25,000 BTU per hour. Running full out for 24 h, then, a home furnace would provide about 600,000 BTU or, in another commonly used unit, about 0.01 tonnes oil equivalent. A gallon of regular gas has an energy content of about 114,000 BTU. Thus, a car getting 20 miles per gal and driving about 10,000 miles per year would consume about 22,800 million BTU or about 570 tonnes oil equivalent.

It is estimated [134] that world energy usage in 2011 was about 12 billion tonnes oil equivalent. Of this, the USA used about 18%, Europe and Eurasia about 24%, and Asia Pacific about 39% (of which China accounted for about 21%). Of the 12 billion tonnes oil equivalent. consumed worldwide, about 4 billion (33%) was oil, 3.7 billion (30%) was coal, 2.9 billion (24%) was natural gas, 0.8 billion (7%) was hydroelectric, and 0.6 billion (5%) was nuclear. Renewables (solar and wind) were negligible (less than 2%). Table 13.1 provides energy conversion factors among various energy units.

S.O. Dean, *Search for the Ultimate Energy Source: A History of the U.S. Fusion Energy Program*, Green Energy and Technology, DOI 10.1007/978-1-4614-6037-4_13,

Table 13.1 Energy conversion factors

1 Btu = the amount of energy needed to raise one pound of water by 1°F
1 million Btu = 0.025 tonnes oil equivalent
1 gal of regular gas = 114,000 Btu
1 kW h = 3,412 Btu

Oil

Because of its widespread use in transportation, a primary concern has been the production, distribution, and consumption of oil. It is estimated [134] that, in 2011, 85 million barrels of oil were consumed daily (31,000 million per year) worldwide, of which the USA consumed about 21%, Europe and Eurasia combined consumed about the same, and Asia Pacific consumed about 32%. Although the Middle East is the largest producer of oil, it only consumes about 9% of the total consumed each year. The total "proved reserves" of oil in 2011 are estimated to be about 1,600 trillion barrels or about 50 years of supply, though additional reserves are expected to be proven in the future. Hirsch notes [135], "Differentiating between oil resources and oil reserves is essential; they are very different. Big new oil discoveries are important, but they will not have the large impact the numbers might imply." The current small estimates of proven reserves, combined with growing populations, emerging economies, geopolitical fears, and concerns about global warming and climate change due to burning fossil fuels, result in mounting pressure to find alternatives to the use of oil in transportation.

Electricity

One of the greatest contributors to increased standard of living in the twentieth century has been the development and growth of electrification. Electricity production and distribution brought a revolution not only in lighting but also in the invention and use of a wide variety of devices that lifted great burdens from humankind and increased productivity. The per capita use of electricity is a primary differentiator between the "developed" and "developing" countries of the world. The amount of electricity per capita available in the "developed" countries ranges from about 1.6 kW in Western Europe to 3.2 kW in the United States. In China, Egypt, and India, on the other hand, it is about 0.2 kW per capita [136].

A typical US home might use about 10,000 kW h of electricity per year, which is about 34 million BTU or 1 tonne oil equivalent. World net electricity generation is about 20 trillion kW h per year or about 1,700 million tonnes oil equivalent. Of this world total, about 66% is generated by the burning of fossil fuels, primarily coal (about 40%) and natural gas (about 25%). Hydro and nuclear fission generate about 16% each; about 2% is generated by geothermal, solar, wind, wood, and waste combined.

Coal

Coal currently provides about 30% of world energy consumption. In 2011, world coal consumption was 3,724 million tonnes oil equivalent, of which the USA consumed 502 million tonnes oil equivalent or 13%. There is an estimated 861,000 million tonnes oil equivalent of coal "proven reserves" in the world, of which the USA has about 28%. The only other countries with large reserves of coal are the Russian Federation (18%), China (13%), Australia (9%), and India (7%). At the current rate of consumption, there would therefore be about 230 years of coal available from proven reserves.

Natural Gas

Natural gas is often used for the generation of electricity instead of coal. It is easier to transport, and gas turbines can be constructed faster and in smaller unit sizes than coal plants. Natural gas is also used instead of gasoline in some vehicles.

Natural gas provides about 24% of world energy consumption. In 2011, natural gas consumption was 3,200 billion cubic meters, of which the USA consumed 690 or 23%. There is an estimated 208 trillion cubic meters of proven reserves [134], of which the USA has about 4%. The Middle East has the largest amount of proven reserves (about 38%), with the Russian Federation the only other country with large proven reserves (21%). At the current rate of consumption, there would therefore be about 55 years of natural gas available from proven reserves. As with the case of oil, additional discoveries are expected, though at higher cost.

Hydro

Hydro (energy derived from falling water) is used almost exclusively for the production of electricity. It is generally agreed that most of the world's potential hydropower has already been exploited. Hydroelectricity provides about 6% of world primary energy consumption. In 2011, energy provided by hydroelectric sources was about 800 million tonnes of oil equivalent

Nuclear

Nuclear fission, energy released from the nuclei of atoms when heavy elements like uranium are split, is also used almost exclusively for the production of electricity, though it is also used for propulsion in naval vessels. Nuclear reactors provide about 5% of world primary energy consumption.

Uranium reserves are not well known and vary in quality. The correlation between reserves and nuclear power plant usage is also difficult to rely on because of the variety of nuclear plant technologies that can use varying grades of uranium. Today's nuclear plants require enriching the uranium ore so that it contains several percent of the more easily fissioned isotope U-235. Estimates are that perhaps 50 years supply of fuel is available for today's type of nuclear power plant. Various other types of nuclear plants have been designed and partially developed that make more efficient use of fissionable elements and could extend the fuel available for several centuries.

When fission occurs, many of the resulting "fission products" are radioactive and consequently are both biologically hazardous and require long (thousands of years) storage as waste. The USA thus far has been unable to agree on how to dispose of long-lived nuclear waste. Other countries alleviate this problem by "reprocessing" spent nuclear fuel, reusing some of it in nuclear power plants, discarding nonradioactive materials, and storing a much lower volume of long-lived waste.

Renewables

Renewables are here taken to include wood, vegetation, geothermal, wind, solar, and waste. Taken together, they provide less than 2% of world primary energy consumption. Wind and solar, together, provide less than one-half of 1% of world primary energy consumption. In principle, and for all practical purposes, the reserves of these forms of energy are infinite. Tapping these sources to meet a significant fraction of world energy demands, however, remains problematical.

The 88,000 tW-year/year of solar energy reaching the Earth's surface is about 9,000 times the amount of current total primary energy use. So, tapping a relative small fraction of this energy could have a large impact on primary energy supply. The energy is very diffuse, however, and not always available to be collected.

Climate Change

The burning of coal, oil, and natural gas currently provides about 87% of world primary energy consumption. However, its use releases carbon compounds and other chemicals into the atmosphere. This degrades air quality and is forecast to lead to "global climate change" with potentially disastrous effects on life as we know it today. Environmental restrictions are becoming increasingly stringent, resulting in economic penalties for using these fuels.

About 6,000 million tons of carbon dioxide are emitted into the atmosphere annually by human use worldwide. The USA is responsible for about 23% of the total. The fastest way to drastically reduce these emissions would be through the construction of nuclear power plants to produce electricity and the concurrent introduction of electric vehicles. To totally replace the current 87% of world primary energy consumption due to coal, oil, and natural gas by nuclear power would require a 17-fold increase in nuclear energy production. Currently, there are about 450 nuclear power plants operating in the world (104 in the USA). Consequently, a 17-fold increase would require the construction of over 7,000 new nuclear plants.

Supply and Demand

Energy supplies are currently meeting demand, although the prices for primary fuels are rising as demand stresses the supply chain. Fuel resources are not equally distributed around the globe, a fact that contributes to economic and military tensions. Energy consumption is also unequally distributed around the world, contributing to great disparities in standards of living and widespread poverty in many parts of the world. Demand is growing rapidly, especially in the "developing" nations, like China and India. This is partly due to meeting the desires of the current populations for increased standards of living and partly due to rapid increase in population relative to the "developed" nations. Currently, about 2 billion of the world's approximately 7 billion people are without electricity. The World Bank estimates that upwards of $4 trillion will be needed to meet the electricity needs of developing countries over the next 30 years.

Although the USA has only about 5% of the world's population, it accounts for about 20% of world energy consumption. Per capita, US citizens consume about 9 times more electricity than the average for the rest of the world and about 15 times more total energy than the typical developing country. The USA spends about $450 billion annually for energy.

Technology advances are slowing the advent of the inevitable energy "crisis," that is, the time when supply can no longer meet demand or cannot meet it at an affordable price.

Perspectives 2012 **14**

Let no man say it cannot be done. It must be done and we have undertaken to do it.

—Franklin Delano Roosevelt, 1942

This chapter presents the personal perspectives of several persons who have played key roles in fusion research and development over many decades.

Charles C. Baker

Is the USA Serious About Fusion Energy?

The United States has been involved in research activities for the realization of fusion energy for several decades. However, we have not been serious enough in the USA in pursuing the potential promise of fusion energy.

Progress in one essential aspect of fusion energy, namely, the development and maturation of plasma physics (for both magnetic and inertial approaches), has been excellent. This has been accomplished by efforts in the USA and throughout the world. International collaboration is and has been a hallmark of plasma and fusion research throughout all its history.

An essential ingredient to progress in plasma physics has been the development of the engineering knowledge and new technology to support the ongoing and, to some degree, the next steps in particular magnetic and inertial fusion energy approaches. However, insufficient progress has been made on the nuclear, materials, and systems sciences needed for fusion energy. To some degree, this has been due to lack of funding but also to misplaced priorities.

S.O. Dean, *Search for the Ultimate Energy Source: A History of the U.S. Fusion Energy Program*, Green Energy and Technology, DOI 10.1007/978-1-4614-6037-4_14,

In the early years, almost all the effort was on plasma physics as it should and had to be. But in the more recent years, particularly the last two decades, the magnetic fusion energy (MFE) program has been generally portrayed as a science program with a potential open-ended future energy application. The MFE program is managed by the Office of Fusion Energy Sciences within the Department of Energy's (DOE) Office of Science. The inertial fusion energy (IFE) program is embedded in the nuclear weapons program in DOE. In both cases, the offices in charge of the program do not have as their top priority developing fusion as an energy source.

The time has come to manage fusion research and development as an energy program and to establish a single office in the US government for managing together the development of MFE and IFE as an energy source. Of course, this program will continue to do cutting-edge science of general value to the nation.

Besides the handicap of trying to develop fusion energy by offices not suited for the task, the USA does not have a well-developed strategy for the development of fusion energy. Such a strategy needs a clear energy mission with a detailed roadmap. *The mission should be to build a fusion demonstration plant in the USA by the year 2040.* The secondary mission is to prepare US industry to design and supply the components for a fusion energy system.

The lack of a clear US strategy does not come from a lack of planning studies. On the contrary, the fusion community has carried out a large number of such studies over many years filling many bookcases with excellent reports. They have time and again carefully examined the technical issues and proposed a wide variety of expanded and new research topics. The lack of a strong US strategy does not result from a lack of understanding of what needs to be done. However, the collection of such studies done by advisory and ad hoc groups does not constitute a viable strategy.

Besides the need for a comprehensive strategy and detailed roadmap, the USA needs the means to implement the strategy. The best means for doing this is to *establish a national fusion energy systems center.* Such a center would coordinate the research and development needed to build a fusion demonstration plant (Demo), objectively evaluate cost options for the Demo, and carry out development path analyses. All this would be done using a systems engineering approach.

As noted above, the best approach will be to develop a strategy and roadmap, implemented through a national center, where MFE and IFE are treated as a single entity. A few past community-based efforts have brought the two activities together with encouraging results. What the USA needs in the future is a single fusion energy program, not two separate programs managed by different offices.

Developing fusion energy is an international effort that needs the best capabilities of the world's fusion programs. An essential element of the US strategy is to recognize that the USA will not be able to develop fusion on its own. It will be necessary to realistically identify those selected areas where the USA should lead, where the USA will contribute but not lead, and where the USA will do little if anything.

One of the impediments, probably the largest, to the development of fusion has been the large and very costly development steps that precede the Demo. This is particularly true for the mainline concepts for MFE and IFE. For one such concept, the tokamak, this is being addressed through an international project—ITER. *An essential strategy for fusion must include development of concepts that lead to cheaper development steps and perhaps smaller fusion energy systems.*

Charles C. Baker is a former director of the fusion energy program at Argonne National Laboratory, a leader of the US fusion technology programs at Oak Ridge National Laboratory, General Atomics, and University of California at San Diego, and a former head of the US ITER Home Team

N. Anne Davies

It was February 1974 when I came to the fusion program. I arrived just in time for the Fusion Power Coordinating Committee review of two proposals for the first tokamak to burn DT. Within 2 weeks, we had approval to put TFTR in the budget request to Congress. Within 2 months, we had also decided to seek funding for two nearer term facilities, the Doublet III at General Atomics and the Poloidal Divertor Experiment at the Princeton Plasma Physics Laboratory. These were heady times for fusion proponents. Within 2 years, the budget increased by an order of magnitude, tokamak physics results were encouraging, and the program had embarked on the development of the technology that would be required for fusion power. Facilities were put in place to develop materials, superconducting magnets, neutral beam and RF heating systems, tritium handling, and pellet injectors. And research on alternate approaches to magnetic fusion was moving forward as well.

Challenges came quickly enough, from the administration, the Congress, and the three construction projects. The administration challenged the need for TFTR. The Congress declined to continue the funding increases, as proposed, and each of the projects, in its turn, faced technical and management problems. The office tackled each challenge and, with the help of the fusion community, dealt effectively with each one. But the program's reputation was tarnished, and we struggled to contain and pay for cost increases. Skeptics and opponents of the program would point to these as management failures and reasons to limit program funding for many years to come.

In the early 1980s, the director of the Office of Energy Research, now the Office of Science, established a federally chartered advisory committee for the program, a committee that would meet in public and advise the director rather than the head of the fusion office. Originally called the Magnetic Fusion Advisory Committee, it is now known as FESAC, Fusion Energy Sciences Advisory Committee. Made up mostly of members of the fusion community, with a few outside members, these committees have, over the years, reviewed the program and provided recommendations on every aspect of it, including priorities and research needs. It was this committee that ultimately recommended the closeout of the magnetic mirror program, although the administration had already taken this action. And it was this committee, after two major budget reductions in the mid-1990s, that recommended a restructuring focused on fusion science and the closing of the highly successful TFTR in order to continue other programs. Again, this was the committee that, after the USA had withdrawn from the ITER project and been on our own for a few years, recommended rejoining that international project to build the first experiment to produce a burning fusion plasma.

In addition to this dedicated advisory committee, there have been myriad fusion review committees, from the secretary of Energy's Advisory Board, a separate Fusion Policy Advisory Committee chartered by the secretary, Congress's Office of Technology Assessment, the president's Committee of Advisors on Science and Technology, the National Academy of Science, and a panel on priorities among energy research projects. These other committees were primarily made up of people outside the program. The fusion program is arguably the most reviewed program in the federal government, and every review has found the science to be of high quality, the promise of fusion worth the cost, and the program deserving of additional funding.

Yet, in the FY 2013 president's budget, the domestic program is devastated in order to provide part (not even all) of the US obligation to the ITER project. Future years' funding looks even bleaker. So while fusion offers the possibility of an energy source cleaner than fossil, safer than fission, with a system that, unlike renewables, could be deployed anywhere and work continuously, and with a virtually unlimited supply of fuel, the prospects for fusion energy development in the USA seem small. When Charlie Baker asks "Is the U.S. serious about fusion," I am afraid the answer is no, we are not.

Perhaps sometime in the future, we shall buy our fusion power plants from one or more of our current ITER partners, who seem determined to develop fusion, with or without the USA. It would not be my preference, but it would be better than not having fusion power at all.

N. Anne Davies joined the magnetic fusion office of the AEC as a staff physicist, ultimately serving as deputy and then director of the Fusion Energy Sciences program at the US Department of Energy. She is a fellow of the American Physical Society and twice was the recipient of the Presidential Rank Award for Meritorious Service.

William R. Ellis

I've thought about what I would like to say in this brief perspective, and I think it is this: to help your readers understand how shortsighted some of our government's high-level decisions have been in science and technology (S&T) areas, including fusion, and what negative effects some of these decisions have had on industry and on the researchers involved. In fusion, consider the US magnetic mirror program. The last big mirror machine to be built in the USA was MFTF-B. I was involved with this project at the Department of Energy, from its inception through construction. I will not comment as to whether the machine was a wise investment or not. But I believe that, having committed to it and funded it, it should have been turned on and used for research. We would have gained valuable new information about magnetic mirror systems and plasma physics as a result. We do not know what we would have found out, but we missed a unique opportunity, not likely to be repeated, to advance plasma and fusion science in this area. There were many reasons given as to why MFTF-B was mothballed after it was built—budget is usually cited as the culprit. But the end result was the same—a major facility was closed down before its job was done—or in this case, even started. It seems this situation has become endemic in fusion and if one thinks about it, in other nationally important S&T initiatives as well. I will mention here only two examples: the SSC and the NASA space program. The Superconducting Super Collider (SSC) was supposed to position the USA to be the world's leader in high-energy particle physics research for decades to come. Instead, it became, as is well known, an expensive earth moving exercise: digging out a big hole and then filling it up again. Since the US government canceled the SSC, the European Union (EU) has taken the lead in high-energy physics experimental facilities worldwide—and seems likely to keep it. The USA has a role there, certainly—as a participant—and I am sure our scientists are contributing significantly. But as far as industrial participation goes, it is EU's high-tech industries that have benefited from this shift and not ours. The NASA space research program has similarly fallen on hard times for funding. One has merely to read the newspapers to see what other countries are doing as they move to fill the gap we leave. Those countries are now developing their own space research facilities and programs. The US industrial base will again inherit the negative consequences of being much less involved than they have been in the past.

The decline of US industry participation in the fusion energy program should be a matter of great concern to all of us. It wasn't supposed to be this way. The Magnetic Fusion Energy Engineering Act of 1980 stated that the Act was intended "to promote the broad participation of domestic industry in the national magnetic fusion energy program." But—as Ed Kintner wrote in his letter of resignation as director of DOE's magnetic fusion energy program office when he concluded the Act would not be implemented—"the plan to increase industry involvement (in fusion has been) postponed indefinitely, and the industrial and economic benefits of high technology spinoffs, surely an increasingly important by-product of an accelerated fusion technology, will be lost."

It seems clear that the world must and eventually will develop fusion energy—at some point in the future. What are the alternatives? If fusion is eventually to become a commercial reality, at some point, industry must have the knowledge and capability to provide fusion technologies. Further, industry provides the link to the customer, which is the electric utilities. My perspective, in summary, is that the federal government has shown itself to be an unreliable partner in supporting high-tech, nationally important, S&T programs over the long term, contrary to what many of us expected—and contrary to what, in fact, we were told to expect by the government (National Defense possibly excepted). Perhaps it is time for someone to document when we began to lose our commanding position, not just in fusion but in other high-tech areas as well. Maybe, it will take some kind of national high-level report clearly stating this, and proving it, to get the government to rethink its funding responsibilities in these areas, and what the huge negative consequences of its S&T funding shortfalls will be—for our national laboratories, universities, high-tech industries and our future.

William R. Ellis is a former head of the US government's magnetic mirror and bumpy torus programs, associate director of research at the US Naval Research Laboratory, vice-president at Ebasco Services and Raytheon, and chair of the ITER Industry Council and the ITER Industrial Consortium.

Richard D. Hazeltine

Fusion Science and Fusion Progress

The campaign to achieve fusion power production is characterized above all by the huge size of its necessary first step. No one has found a tabletop prototypical fusion reactor that proves the concept, much less a route toward short-term economic reward, and these circumstances have burdened the program from its beginning. But fusion research is also characterized by rich scientific interest, with an illustrious history of scientific discovery and multiple ties to other disciplines. In fact research supported primarily by the fusion program has engendered new scientific fields of considerable and growing consequence; plasma astrophysics and high-energy-density physics are two obvious examples. The thesis of this essay is that a broad menu of scientific research is an essential component of the fusion quest, one that carries particular weight as the program's increasingly urgent ambitions conflict with shrinking budgets. (Full disclosure suggests pointing out that this thesis reflects its author's background as a plasma theorist in a university faculty.)

The dichotomy between scientific breadth and programmatic focus is closely related to other fusion research conflicts, including "science versus energy" and "innovation versus main chance (tokamak)," topics that have occasioned many community debates. However repetitive and even dreary these debates have become, the issues are real. In particular, the question of breadth sharply affects the level of contact between fusion science and other disciplines of science and technology; examples are solar and magnetospheric physics, high-power laser development, and fluid turbulence (theory and experiment).

The potential rewards accruing from contact between neighboring scientific disciplines are enormous and widely appreciated in the broader scientific community. For example, the physics of condensed matter, including superconductivity and 2-D Hall effects, has had extremely fruitful interaction with elementary particle physics. Fusion examples include the value of turbulence theory in understanding the relation between velocity shear and tokamak transport, or the importance of Hamiltonian theory in understanding flux-surface integrity and magnetic diffusion. An isolated, parochial research program is an impoverished one.

There are also indirect benefits of colloquy between fusion science and its intellectual neighbors. In particular, the program becomes especially vulnerable to funding disasters when its goals and scientific quality are not broadly appreciated throughout the scientific and technological community.

An often-mentioned advantage of fusion program breadth concerns workforce development, an increasingly critical issue in US fusion research. Attracting talented, dedicated people into the fusion quest is a particular responsibility of the academic community, and the academic lesson is clear: students and young scientists are attracted to a program that, while keeping the power-production objective in view, can also support research in areas off the critical path. This diversity brings obvious intellectual appeal but is attractive as well for more practical reasons. Students now entering the university know that, at about the time when they begin to consider retirement, either fusion power will be joining the grid or any sizable search for fusion energy will have been abandoned. In either case, research support for fusion will be much reduced from its present levels, and a major practical impetus for studying the science underlying fusion will be lost. These losses become manageable, if not harmless, if the student's training and research experience has been sufficiently broad.

In summary, the vitality of fusion research has from the beginning depended upon a diverse, multifaceted research agenda. As the program begins to approach its energy-production target, such diversity seems more important than ever.

Richard D. Hazeltine is a professor of physics at the University of Texas, Austin, and was chair of the US Department of Energy's Fusion Energy Sciences Advisory Committee (FESAC) and chair of the American Physical Society Division of Plasma Physics.

Robert L. Hirsch

A Fusion Failure

Tokamak fusion will almost certainly fail to become a viable, commercial electric power system. What happened?

Over 50 years ago researchers came to realize that fusion plasma physics was extremely complicated, and it took decades to develop a substantial physics knowledge base. After various early magnetic fusion concepts showed poor promise, the tokamak concept emerged as a bright spot, and the world fusion community increasingly focused on that concept as a promising option for practical fusion power. If tokomak physics could be successfully developed, it was assumed that the requisite engineering would evolve thereafter. As it turned out, engineering considerations of tokamak fusion turn out to be commercial showstoppers.

The cornerstone of world fusion research today is the inordinately expensive ITER tokamak project under construction in France. All of the government-sponsored magnetic fusion "eggs" are in that "basket" in popular perception. The broad-based assumption is that ITER's successful operation will lead directly to a similarly configured commercial fusion power system. A few of the many reasons that assumption is almost certainly incorrect include the following:

- ITER-like fusion power plants will be extremely large and complicated, requiring large volumes of finished materials, which are inherently very expensive. Low cost is a major requirement of electric power generation.
- DT fusion inherently produces neutrons, which induce radioactivity in structural materials. The resulting radioactivity will require complicated remote operations, expensive maintenance, and the disposal of large volumes of radioactive waste. DT fusion is not "clean," as often claimed.
- Contrary to popular belief, ITER tokamak fusion reactors will not be "inherently safe," because tokamak plasma containment requires very large, high-field superconducting magnets, which contain large amounts of stored energy, which could be explosively released, if one or more of the magnets were to suddenly go normal. While this may seem like a low probability event, it is nevertheless possible.

Consider what just these three factors mean in practical terms. The large size of an ITER-like tokamak fusion reactor means huge cost. The fusion community should have been searching for an inherently small-sized concept, which might have a chance of being low cost. As currently understood tokamak physics indicates, scaling to extremely large power outputs could trend toward reasonable economics, but that means multigigawatt electric power plants, which utilities do not like, especially when new, complicated technologies are involved.

The fact that DT tokamaks inherently produce radioactivity means that in the USA, the Nuclear Regulatory Commission (NRC) will have responsibility for licensing such facilities. Based on fission reactor experience, NRC will require extensive and expensive safety features. Among the most obvious will be a blast-resistant building to contain the hypothetical large energy release associated with normal-going superconducting magnets. A commercial ITER tokamak building might then be similar to what is required for fission reactors. Because tokamaks are so very large, such a building will dramatically increase overall facility cost.

How did this physics/engineering technology imbalance happen? One view is the belief by researchers and management that fusion research and power plant engineering could be done in a serial manner, which would allow a focus on the physics first, followed by the engineering of a commercially viable power system. In fact, that is not a correct assumption. In my commercial energy R & D experience, I repeatedly saw that engineering and market realities must be integrated and balanced early in program development, if an attractive idea is to have a chance of succeeding. Had such a balance been established in magnetic fusion research decades ago, it is almost certain that the tokamak approach would have been channeled in a different direction or discarded.

A serious, independent, unbiased commercial evaluation of the ITER tokamak approach to commercial power is lurking and will happen. The result will almost certainly be a very negative conclusion and a collapse in financial support, possibly with a condemnation of the viability of magnetic fusion. Bottom line is the following: The magnetic fusion community worldwide is likely in for a very bad experience.

The jury is still out on inertial "little bomb" fusion concepts. At this writing, the folks involved with those concepts are still working toward breakeven, which will hopefully happen in the near future. Thereafter, when researchers seriously embark on the goal of commercial fusion power, they will hopefully benefit from the experience of the magnetic fusion program.

I remain hopeful that commercial fusion power will someday become a reality. Recall one of the explanations for the "Sherwood" code name for the fusion program when it was classified in the early days: It "sure would" be wonderful.

Robert L. Hirsch was director of the US fusion program and assistant administrator of the US Energy Research and Development Administration (ERDA) during the 1970s. He later was an executive at Exxon, Arco, and the Electric Power Research Institute (EPRI) and lead author of the book "The Impending World Energy Mess" (Apogee Prime Books, 2009).

B. Grant Logan

Taking a Longer, If Slower, Path to Fusion's Ultimate Potential

Fusion energy advocates are generally optimistic types—we almost have to be optimistic to stay in pursuit of the fusion energy dream for so long. Lately, our optimism has been sorely challenged, especially in the USA: real $ funding for US fusion research is still less than half of what it was in the early 1980s after the last energy crisis. The mean age of the fusion workforce is increasing, and many of its researchers are approaching retirement. US magnetic fusion facilities are also aging, and MFE ignition is not expected in ITER before 2026. The commencement of inertial fusion ignition experiments in the US National Ignition Facility in 2009, while achieving unprecedented confinement parameters of rho-r and ion temperature, may require more time to reach ignition than initially planned. The many scientific and technical challenges of achieving practical and affordable fusion energy are well known to both advocates and critics of fusion, and while the annual cost of fusion R&D must more than double to demonstrate net electricity by fusion in 20–40 more years, we know that the R&D cost could be afforded by the world's wealthiest nation, even if fusion R&D were not invested in as much as other energy R&D, especially near term energy R&D.

But perhaps most discouraging to fusion optimists are the challenges coming from many critics of fusion who assert (a) that advances in renewable energy technology, including solar, wind, biofuels, and energy efficiency, will likely suffice, thus obviating any need to develop fusion as a long-term option; (b) that fusion development would take too long anyway to reduce greenhouse gas emissions fast enough to avoid unacceptable global warming climate impacts; and (c) that even in the event green energy growth were too slow to save the planet, nuclear fission energy and natural gas could be ramped up fast enough, and could be designed safe enough, and that fission reactors and/or gas turbines would be much more affordable than fusion plants. DOE Secretary Chu and former Undersecretary Koonin have pointed out that no future energy source, regardless of its environmental and safety advantages, can escape the necessity of projected economic competitiveness (usually followed by quoted examples of advanced solar and wind designs with lower projected unit costs of electricity).

Since fusion is still a long way off in any event, we would be better off to focus on fusion's ultimate potential (i.e., advanced concepts) in this competitive future energy market to motivate at least modest increasing levels of funding worthy of an attractive long-term energy option. Understandably, many fusion advocates are impatient, trying in vain to convince skeptics fusion could make competitive cost of electricity after a first-generation demonstration power plant based on what we could design today if given a crash-program level of funding. To justify a more credible, longer path strategy to fusion energy, it would be helpful to document a "brutally honest" projection of potential energy and water challenges facing our grandchildren joining a world population heading toward ten billion people. Many studies show energy and water would be twin-related issues in the globally hotter planet our grandchildren will most likely have to face. Increased fresh water supply, not just conservation alone, is already a growing need in more countries, and so more energy will be needed to pump fresh water over greater distances and ultimately to desalinate ocean water in regions where every available river and lake are already tapped out.

Against the argument (a), above, rest assured we can and will find much better fusion designs in the future than we know how to build today. It isn't just renewables which can improve; we can innovate fusion in dramatic ways if we put our minds to it, including harvesting high-temperature fusion energy for synthetic fuel/hydrogen production and exploiting direct conversion. These advances will take more R&D time, but then, fusion energy can more likely be unique and competitive. We can dispense with the argument (b) above that fusion cannot be developed in time to stop global warming and start arguing the other side—that fusion, with limitless fuel supply, will be necessary to cope with the huge and growing impacts of global warming in all countries. As for argument (c), to be sure, nuclear fission and natural gas must and will grow to fill the energy gap in the next several decades created by a globally hotter planet, but the scale of new energy demand needed, e.g., up to ~5 TWe just for global ocean desalination alone, will force such a dramatic demand push-up in new reactor start-up fuel cost, and even on natural gas, as to put a constraint on the rate of growth of fission and natural gas power. When such growth rates hit the fuel-supply limits in the future, we should aim to have fusion (with lowest fuel costs) ready to step in to augment the fast-growing power needs of ten billion people who have to live on a hotter planet.

B. Grant Logan was director of the heavy ion fusion program at Lawrence Berkeley National Laboratory and director of the US Virtual National Laboratory for Heavy Ion Fusion.

Robert L. McCrory

Perspectives on Inertial Fusion Energy: 2012

The demonstration of inertial confinement fusion (ICF) ignition on the National Ignition Facility (NIF) [137, 138] over the next few years will represent a major step in ICF research and an important first step on the long path to inertial fusion energy (IFE). Significant science, technology, and engineering developments will be required to meet the goal of producing electricity commercially. There are a number of ICF concepts (in various stages of development) that could prove to be more optimal for IFE than the current indirect-drive baseline [139] of the National Ignition Campaign [140]. As stated in the interim report of the National Research Council's Committee on the Prospects for an ICF Energy System, "It would be premature at the present time to choose a particular driver approach as the preferred option for an inertial fusion energy demonstration plant" [141]. The underlying science must mature to meet the gain and reliability requirements of an IFE plant.

While the nation is on the verge of demonstrating ICF ignition, even if those results were to meet the requirements of an IFE plant, the technologies needed for such a plant have either not been demonstrated or are too expensive for commercial energy production at this time. High-repetition-rate drivers have not been demonstrated at the scale required for a commercial plant nor has their cost-effectiveness been validated. There are many reactor systems issues that need to be understood and optimized. There has been no demonstration of the cradle-to-implosion mass production of IFE targets. New approaches will be required. For example, a collaboration led by the Laboratory for Laser Energetics (LLE) is developing a "lab-on-a-chip" technique to make, fill, and overcoat foam shells that scales to mass production for direct-drive targets [142].

Direct-drive ICF is a leading candidate for an IFE plant [143]. Direct drive has significant advantages, including a simple target design and coupling significantly more energy to the compressed core than with indirect drive. The University of Rochester Laboratory for Laser Energetics (LLE) is the lead laboratory for the development of direct-drive ICF and has made steady progress in understanding the underlying physics over the past decade. This includes the demonstration of areal densities of ~300 mg cm^{-2} cryogenic deuterium–tritium implosions on the OMEGA Laser System at 25 kJ [144, 145]. If the design is scaled in laser energy E_L, the areal density changes approximately as $(E_L)^{1/3}$ [146]. Therefore, the OMEGA results scale to an ignition-relevant areal density of ~1.6 g cm^{-2} for a laser energy of 1.5 MJ. LLE plans to perform direct-drive-ignition experiments on the NIF in the next decade, employing the polar-drive concept that allows spherically symmetric direct-drive implosions, while NIF is configured for indirect drive (cylindrical but not spherical symmetry) [147]. Two-dimensional hydrodynamic simulations predict that a gain of 32 can be achieved with the expected levels of target and laser nonuniformities [148].

The path to inertial fusion energy can begin in earnest after ignition is demonstrated on the NIF in the next few years. Direct-drive ICF is a leading option for an IFE power plant. The road will be long and expensive!

Robert L. McCrory is vice-president and vice provost of the University of Rochester and director of the University's Laboratory for Laser Energetics. He is also professor of physics and professor of mechanical engineering at the University.

Dale M. Meade

Time for the Fusion Community to Focus on the Future

From the early 1970s to mid-1990s, the US fusion program aggressively pursued the mission of achieving fusion energy and made dramatic advances in fundamental scientific understanding of fusion plasmas as well as extending the frontier of laboratory plasma parameters to fusion plasma conditions. This progress was enabled by a national commitment in the mid-1970s to support fusion energy research and, most important, a community focus on the goal of fusion energy with a "yes, we can" approach that found solutions to problems encountered along the way. Since the budget collapse in the mid-1990s, the US fusion community has lost its focus on the fusion goal, slipped into a "No, you can't" approach that dwells on problems instead of finding solutions. As a result, progress toward fusion has stagnated, and the US has fallen from a world leadership position in fusion research to that of a supporting partner in ITER, while the goal of commercial fusion energy recedes 1 year per year.

Since the late 1980s, the USA has been technically ready to take the next major step—we now know that the proposed Compact Ignition Tokamak (CIT) would have ignited ($Q=35$), but the community bowed to the "No, you can't" naysayers in the Department of Energy and lost the opportunity to produce a burning plasma with about 300 MW of fusion power by 2000. That burning plasma mission is now taken up by ITER, which is scheduled to produce a burning plasma with 500 MW of fusion power in 2027. While many feel this is not the optimal approach, we must continue the international momentum that has been established to construct a power-plant-scale fusion device.

It is widely acknowledged that the world fusion program needs to be broader than just ITER. Indeed, Europe and Japan have formally agreed to a broader approach that includes a $1B-scale superconducting non-burning plasma experiment, JT-60SA, a Demo design center, and the design of a neutron-materials irradiation facility. A number of the participants in ITER have made major commitments to increase their domestic fusion program by building significant non-burning plasma facilities including the Republic of Korea with the KSTAR superconducting tokamak almost identical to TPX, the device the USA was unable to find the resources to build in the mid-1990s. China is now operating EAST, a superconducting tokamak, the scale of DIII-D, the largest US fusion research facility, and Europe continues to extend their world leadership position in fusion with a major upgrade of JET, construction of a $1B-scale optimized stellarator W7-X in Germany, and continued operation of medium-scale tokamaks.

Several technical reviews have identified the fusion nuclear science issues that must be addressed to proceed with a demonstration of fusion energy including long-pulse burning plasma sustainment, efficient particle and heat exhaust from the fusion plasma, closing the fusion fuel cycle, development of materials suitable for the fusion environment, and operational experience with fusion systems. These issues will not be resolved by ITER and the non-burning facilities described in the preceding paragraph. The need exists for a fusion nuclear science program to address these issues. Such a program will be at least as challenging as the attaining fusion plasma condition that we have pursued for the past 50 years. Now is the time to begin to address the nuclear science issues, and the opportunity exists for the USA to take a leadership role in this area.

The US fusion community should put forward a goal-oriented plan to address the critical fusion nuclear science issues that will increase the credibility of fusion energy. Executing this plan will require refocusing the fusion community on the goal of achieving fusion energy on a shorter time scale.

Dale M. Meade was leader of the Tokamak Fusion Test Reactor (TFTR) at Princeton Plasma Physics Laboratory (PPPL), deputy director of PPPL, and proponent of constructing a Fusion Ignition Research Experiment (FIRE).

John H. Nuckolls

Inertial Fusion Energy: Super Lasers and Super Implosions

In spring of 1960, I calculated a *fusion microexplosion*, a 50 MJ high-gain microexplosion, initiated by a tiny spherical implosion in a small *hohlraum*, heated by a multi-megajoule pulse of radiation energy. A few months later, when Maiman announced the first successful laser, we began to analyze the feasibility of heating plasmas to high temperatures, laser-driven implosions, MJ super lasers, laser fusion, and weapons physics applications. In 1961, I calculated use of temporal pulse shaping to create super implosions which initiate low-cost, propagating microexplosions required for practical fusion power applications.

Since 1963, the Livermore Laboratory has been funded by the Department of Energy nuclear weapons program to develop a series of larger, more advanced super lasers, including the 20-beam, 10 KJ Shiva laser in the 1970s and the 10-beam 30 KJ (blue light) Nova laser in the 1980s. In 1972, our calculations of super implosions directly irradiated by lasers were declassified; later, our indirectly driven (hohlraum) super implosion targets were declassified. In the 1990s, when underground nuclear testing ended, the USA launched the Stockpile Stewardship Program, including design and construction of the 192-beam, 1.8 MJ National Ignition Facility (NIF). NIF was dedicated in 2009.

National Ignition Campaign (NIC)—As of August 2012, the very successful NIF super laser is focusing 2 MJ, 500 TW temporally shaped pulses of blue laser light into cm-scale hohlraum targets, energizing millimeter-scale super implosions designed to ignite propagating 20 MJ fusion microexplosions. Experts are recording volumes of experimental data from NIF's array of several dozen advanced diagnostic instruments, analyzing this data with powerful supercomputers and codes (billions of times more powerful than in the 1960s) and improving the super implosion.

Outstanding progress has been made. The imploded central core of deuterium–tritium (DT) has achieved ideal ignition temperatures and is within a few-fold of the required ignition pressure. The surrounding DT shell is several thousand times liquid density—within a factor of two of the required density. To achieve ignition, the hot core density–radius product must be increased twofold to absorb the DT alpha energy and enable thermonuclear bootstrapping.

Challenging difficulties have been encountered. Hydrodynamic instability growth and mix are larger than calculated, and implosion efficiency is significantly lower. Target improvements are required. Black swans cannot be precluded. More NIF experiments are required to demonstrate that directly driven super implosions can achieve higher implosion efficiency while avoiding excessive growth of laser and plasma instabilities.

Target designs with higher implosion efficiency, but with reduced implosion requirements, may be feasible, e.g., tamped non-ablative designs or designs with magnetized, preheated DT. Practical fast ignition targets (which utilize a petawatt ignition laser) may be invented to minimize laser energy, relax super implosion performance requirements, and enable enhanced thermonuclear propagation to achieve very high gains.

Fusion Power: Major Economic and Technological Challenges—Systems studies conclude that practical inertial fusion power plants are feasible. After ignition and high gains are achieved, and several beamlines of an efficient, high-repetition-rate diode-pumped solid state laser (or other driver) are demonstrated, substantial inertial fusion energy funding will be required to construct a burst mode, high-heat-capacity inertial fusion energy system, including a driver, a target factory, and an explosion chamber protected from neutron, X-ray, and exploding debris damage by fluid shielding. Repetition rate could be gradually improved from hours to minutes to seconds—leading to an experimental fusion reactor. Fusion neutrons might also be used to burn nuclear waste.

Government funding of fusion energy development may be limited while the US economy is recovering. A global fusion power system will cost trillions of dollars. Innovations that demonstrate a significant economic advantage may attract massive private sector funding and greatly accelerate deployment.

After fusion driver, target, and reaction chamber costs are minimized, significant reductions in the dominant balance of plant capital cost, and increases in thermal electric efficiency, may require invention of practical direct conversion electrical generating systems and deuterium burning targets.

John H. Nuckolls is a pioneer of the field of inertial confinement fusion and is director emeritus of the Lawrence Livermore National Laboratory.

Richard F. Post

Magnetic Confinement Fusion Power Research: Thoughts After 60 Years of Involvement

In 1951, Herbert York gave three lectures at the Lawrence Berkeley Laboratory on the classified controlled fusion research then in progress at Los Alamos, led by James Tuck, and at Princeton, led by Lyman Spitzer. Intended to stimulate interest in the upcoming Livermore Lab, York's lectures did the trick, and soon I joined the new Lab, heading a small team investigating the "mirror" approach to magnetic confinement fusion. Mirror research has been a passion for me ever since. In what follows, I will use items from its history to help explain my feelings about the present state of magnetic fusion research.

Launched with strong support from AEC Chairman Lewis L. Strauss, the first two to three decades of US magnetic fusion research were characterized by breadth, both in pathfinder experiments and in major breakthroughs in theory. I like to think of that period as analogous to the establishment of trail markers and warning signs as one climbs a steep mountain slope. In my view, there were two significant trail markers, one experimental and one theoretical, that emerged early on in our mirror research. The experimental one, the "tabletop" axisymmetric mirror with the field line geometry of a cylindrical "party popper," threaded axially by magnetic field lines that are constricted at each end (the mirrors) and then expanded. In this experiment, magnetic compression of an injected plasma led to the formation of a 2–3 cm diameter cigar-shaped hot electron plasma about 30 cm in length. As shown by measurements on escaping electrons, this plasma was confined stably for many milliseconds in agreement with "classical" mirror end losses. More importantly, during its confinement, its diffusion rate across the confining magnetic field was almost immeasurably slow. In fact, it was at least 100,000 times slower than the cross field electron loss rates then being measured in Spitzer's stellarator, rates that agreed with the turbulence-dominated losses predicted by Bohm. As for the theoretical trail marker, our mirror results were in fact consistent with theoretical studies by Teller and Northrop in which they showed that for an axisymmetric mirror field, the trapped ions and electrons of the plasma should be confined to close axisymmetric drift surfaces, involving no electric currents parallel to the field lines. This situation must be contrasted with that in the doughnut geometry tokamak and stellarator, where the drift surfaces are not closed and where, in order to achieve a confined state, it is essential that electrical currents flow along the circling magnetic field lines, a situation that promotes plasma instability.

While at the time we did not understand what kept the tabletop plasma stable against the MHD-induced cross field drifts predicted by Teller and by Rosenbluth and Longmire, we now believe we have a plausible theoretical explanation (derived by Dmitri Ryutov), together with experimental confirmation of his theory by experiments performed on the axisymmetric gas dynamic trap mirror experiment in Novosibirsk, Siberia.

Until the late 1980s, there was steady progress in mirror theory and experiment, including the seminal concept of the tandem mirror as proposed by Dimov in Novosibirsk and Fowler and Logan at Livermore, followed rapidly by the construction of several tandem-mirror experiments in the USA, Novosibirsk, and Japan.

Then an unfortunate change occurred. In the USA, the decision was made to concentrate virtually all magnetic confinement research, both experimental and theoretical, on the tokamak—the front-runner at that time. It was as if, after the first lap had been run in a horse race, the judges saw which horse was ahead and at that point disqualified all of the other horses. Of course, the tokamak deserves support. However, it should not be at the expense of sacrificing a critical feature, namely, programmatic breadth, of any major scientific undertaking, be it cancer research or fusion research—that addresses a critically important problem for this country and for the world. The question is the following: Has the magnetic fusion program strayed off the surest path to practical fusion power? If so, can it be brought back on track?

Of course it can! I believe attitudes must change, from "tokamak ITER and its follow-on Demo represent *the only sure path to practical magnetic confinement fusion power*" to "the surest path to practical magnetic confinement fusion power is to use the trail markers and warning signs that have been established in the last 60 years to define a broadly based R and D program that supports and welcomes the study of promising approaches." Such a change in attitude is essential if we are to attract bright young scientists and engineers, those who now believe that magnetic confinement fusion based on the tokamak is a "done deal" to which they can make little, if any, contribution. If those changes are made, we not only will restore to fusion research that sense of excitement that we old timers felt when we began the fusion quest but also, I feel, will guarantee its success. Finally, let me say this to the budget-minded doubters: Human psychology being what it is, once the attitude is changed to a positive, welcoming one, it can work its magic at almost any level of financial support. I have seen it happen.

Richard F. Post is a pioneer of the US magnetic fusion program and has been a leader and spokesman for the magnetic mirror approach to fusion at the Lawrence Livermore National Laboratory during a distinguished career spanning 60 years.

Fred L. Ribe

Recollections and Perspective from 40 Years of Magnetic Fusion Research

In fusion research (CTR as we called it in the 1950s), many of us at Los Alamos got our first exposure to plasma theory from Conrad Longmire who had worked with Marshall Rosenbluth and Edward Teller on the thermonuclear bomb. Jim Tuck started CTR at Los Alamos remembering the classified Z-pinch work in the USSR and UK that had led to disappointment when the neutron emission turned out to be from plasma instabilities.

Tuck called the first toroidal Z pinches “Perhapsatrons”. He led a group of experimentalists through various pulsed, high-density concepts. The group included K. Boyer, R. Lovberg, J. Marshall, D. Nagle, J. Phillips, W. Quinn, and G. Sawyer, among others. The concept I settled on was the theta pinch, which was a continued source of satisfactory research. Los Alamos took the Scylla theta pinch (so named by Jim Tuck) and John Marshall’s plasma gun to the 1958 Geneva Conference where we met our Soviet counterparts, including Lev Artsimovich.

Scylla was a good hot-plasma producer, as was the Pharos theta pinch at NRL. Our two groups did radiation physics on our unique million-degree plasmas. At Los Alamos, the hot-ion Scyllas progressed to longer versions. The Moscow tokamak T-3 was confirmed to have sustained hot plasma by the Culham laser-scattering group, and the rush to tokamaks began. However, Los Alamos and Livermore stayed with their Scylla theta pinches, toroidal ZT pinches, and magnetic mirrors. The final Scylla was the toroidal Scyllac high-beta stellarator, and the last mirror was the MFTF-B. Both were terminated as US fusion development concentrated solely on tokamaks and inertial confinement (ICF).

The international ITER tokamak now provides the focus of the world magnetic fusion community’s work toward a power reactor. Of the pulsed high-density experiments of the 1950–1980 period, only the field-reversed configurations (FRCs) derived from the theta pinch remain under development.

A major reason for setting aside the high-density magnetic fusion approaches was the perceived problem of adapting their pulsed plasma formation to a fusion-reactor environment. Field-reversed configurations (FRCs) are being actively pursued. Their pulsed environment remains. This environment is shared by the laser-driven (ICF) concepts. The duty factor of ICF plasma formation is a particular problem.

ITER is advancing toward a large, low-beta, low power density reactor. It is unfortunate that means have not yet been found to advance the high-beta magnetic fusion concepts of the 1950s to 1980s to the reactor regime. Their beta values are an order of magnitude greater than that of the tokamak, and the power densities are two orders of magnitude higher, which would result in smaller fusion-reactor cores. The materials and engineering problems of their reactor embodiments, many of which are shared by the tokamak, need further research. Conceptual reactor studies have already been done.

It is important to have a diversified magnetic fusion program in case the tokamak doesn’t fulfill present expectations. Young researchers coming into fusion need to acquire basic understanding, which can be provided by alternative experiments, particularly those at universities.

Fred L. Ribe was head of the fusion program at Los Alamos National Laboratory, a professor at the University of Washington, and chair of the US Department of Energy’s Magnetic Fusion Advisory Committee (MFAC).

John Sheffield

In 1956, Bas Pease, later director of the UK fusion program, made the following comment, "…Our vision is of a power station, sited perhaps on the coast, with a pipe bringing water from the sea, helium leaving by the chimney and electrical power flowing into the grid. We do not know what to put inside the power station (laughter)…"

We've come a long way since then, baby, with tremendous progress across the board and real prospects for significant gain in IFE and MFE facilities, but we should not forget four important points:

- To be careful how we present fusion in relation to other energy sources.
- With D–T reactors, the challenge of developing radiation-resistant materials may be greater than producing net energy.
- The ultimate fusion fuel is deuterium.
- High availability will be difficult to achieve in a complex fusion reactor.

In regard to energy sources, presently, the world has numerous options, and the primary issues are their distribution and the need to reduce pollution. The situation in which the developed and emerging nations consume most of the readily transportable and cheap fossil fuels presents a huge problem for the developing world: data shows that population growth goes inversely as energy consumption per capita. The value of what will likely be relatively expensive fusion energy is if it leads the developed and emerging nations to reduce their dependence on fossil fuels and make them available to the developing world. In the very long term, the only massive energy resources will be solar in one form or another and fusion—so having fusion energy available will be desirable at any time.

One noticeable trend in magnetic fusion-reactor studies over recent decades has been a steady decrease in their neutron wall loading. I suspect that this is partly because the development of radiation-resistant materials has progressed more slowly than anticipated, and designers are more cautious about assuming how long a blanket might last.

When we talk about fusion fuel being limitless, we are referring to deuterium. Ultimately, a deuterium reactor should be the goal. One possibility that would alleviate the neutron-materials problem would be to remove as much as possible of the tritium produced before it burns. This appears to be a possibility for tokamaks and stellarators and in batch-burning systems such as compact tori and inertial fusion energy. Of course, in the latter two cases, thick liquid lithium walls might provide a solution to the neutron problem.

Finally, I suggest that achieving high availability, i.e., >0.80, will be the greatest challenge for a fusion reactor. The availability of a system depends upon the failure rate of its components and the time to repair or replace them, and there is hardly any data on fusion-reactor systems.

In summary, we should not be too hung up about fusion having to be cheaper than all other sources because its importance to improving global energy resources transcends that one issue; we should pay more attention to materials issues and options that could use tritium-suppressed deuterium; we should put effort into developing options that would be more readily maintainable.

John Sheffield was director of the Fusion Energy Division at the Oak Ridge National Laboratory, executive director of the Joint Institute for Energy and Environment at the University of Tennessee-Knoxville, and chair of the US Department of Energy's Fusion Energy Sciences Advisory Committee (FESAC).

Ken Tomabechi

Use of thermonuclear fusion promises a great potential to contribute to the future world, so that strong efforts should be made in order to realize its commercial uses before the middle of this century in any case.

One of the big advantages of fusion reactors over fission reactors is that fusion reactors do not produce troublesome radioactive materials to be treated such as fission products produced in fission reactors. Major radioactive materials to be dealt with in fusion reactors are tritium fuels and radioactive materials produced by absorption of neutrons generated in the core. The latter can be controlled by proper selection of materials to be used in fusion machine.

In the following discussions, let us adopt here an energy unit of zeta joule (ZJ), i.e., 10^{21} J. The present world energy consumption is about 0.5 ZJ a year, and presently, known recoverable resources of coal, oil, and natural gas are 22.4, 6.7, and 6.4 ZJ, respectively. There was a report that the amount of lithium resources in Western world was 8.3 Mt, which would be able to generate 175 ZJ, implying that fusion will be able to supply energy to the world for hundreds of years. If the amounts of lithium resources in China and others are added, the energy to be produced will become much larger. Furthermore, studies are being made to extract lithium from the seawater that contains lithium at a concentration of 0.17 ppm. It means that the total lithium in the seawater is calculated as 240 Bt and the amount of energy to be produced from that lithium by DT fusion will be 5.1×10^6 ZJ that corresponds to the world energy demands for millions of years to come.

Limitation of availability of tritium fuel that has a relatively short half-life of 12.3 years may constrain a quick and wide deployment of deuterium–tritium fusion reactors at an early stage of its commercialization. Therefore, in order to alleviate such difficulties, efforts should be made to develop fusion reactors having as high tritium breeding ratio as possible, regardless of the type of fusion reactors, i.e., either magnetic confinement or inertial confinement.

It should be noted however that in our study, it seems possible to design a DT fusion reactor to be started up its operation without any external supply of tritium fuel at all, i.e., by generating the necessary tritium fuel in the machine itself. The machine is started up first with pure deuterium plasma in the core. Then, the small amount of tritium produced in the core plasma by the DD reactions and the tritium produced in the blanket by 2.45 MeV neutrons generated by the DD reactions will be extracted and recycled back into the core. Such operational procedures will be repeated many times, so that the tritium concentration in the core plasma will increase gradually and finally reach the designed conditions of the DT plasma. Such a commissioning procedure may require about a couple of months of those special operations while using a supply of the necessary electricity from the connected grid.

Ken Tomabechi, a longtime leader of the international fusion research effort, was the first director of the ITER project and a research advisor to the Central Research Institute of Electric Power Industries in Japan.

Alvin W. Trivelpiece

There is a natural pace at which humans advance their body of accumulated knowledge. If an infinite amount of money was made available to Galileo, he could not have invented and developed a device to release energy based on the fusion of certain low mass atoms. Even Leonardo da Vinci might have had trouble with this invention. In some sense, this is true today. An infinite amount of money could not produce a commercially viable fusion power system delivery tomorrow. On the other hand, our current body of knowledge informs us that our Sun's fusion reactor has produced a steady long-term source of energy for planet Earth. We all hope that this controlled fusion reactor will last for a long time.

The invention of nuclear weapons demonstrated that it is possible to release enormous quantities of energy through the controlled thermonuclear fusion processes. Is there an intermediate regime between nuclear weapons and the Sun that involves laboratory-scale controlled thermonuclear fusion reactions? I believe the answer to this question is yes. So much for my perspective!

Even so, efforts to realize the goal of developing a device that can produce such controlled thermonuclear fusion reactions at a practical scale that would meet the requirements for reliability, safety, and environmental acceptability remain to be achieved.

For magnetic fusion, there is a long history of sequential experimental and theoretical research and development. This has resulted in progress toward the goal of achieving conditions leading to the release of more fusion energy than was required to create those conditions. I am not aware of any analysis that proves this cannot be done, but even if it might be possible, it may not be commercially competitive with other sources of energy in the immediate future.

For some time, there has been a lively debate about which approach and technology to achieve magnetic fusion is the correct one. The 1983 Versailles G-7 Summit established the Technology Growth and Employment activity. Paolo Fasella and I were the cochairmen of the fusion subgroup of this activity. We agreed that we should produce a long-range plan that identified the sequence of realistic steps required to make magnetic fusion work without regard to which country did which task. This may have been a bit naïve, but it did identify a progression of steps that were practical. At that time, there was a consensus that the available talent was a more critical commodity than funding. That is, it did not seem a good idea to have each nation duplicate the experimental facilities, when a collaborative approach would benefit all nations and take good advantage of the available talent, while at the same time reducing the cost. Given the nature of the problems that required solutions, this collaborative approach would not give any of the involved nations an unfair competitive or commercial advantage.

Eventually, the Soviet Union was included as a team member, and there was an agreement that led to the International Thermonuclear Experimental Reactor (ITER). This was originally conceived during a period when all the parties were financially stable and could make a contribution to this effort that would sum to about $4 billion.

Unfortunately, those conditions didn't last, and several of the partners had financial or political problems that resulted in less than full participation. Further, some disagreements emerged that led to the present situation with a substantially higher cost ITER that is being built in France.

My current circumstances have not provided me with sufficient detailed information regarding the current status of ITER, but the project has been delayed in its goal of demonstrating that sustained fusion is possible and now focuses on one involving a level of engineering to make a facility that might generate significant levels of energy. From my perspective, the goal should remain one that seeks to prove that magnetic fusion can work. In that regard, the effort should be funded at a level that answers the key questions in such way that the rate of funding leads to the lowest total cost. Funding at a level that doesn't answer the critical questions leads to greater cost than one that makes the cost versus time curve such that the answer is obtained at lowest total cost. Unfortunately, we seem to have past the point where such consideration is still possible.

Regarding Inertial Confinement Fusion (ICF), I am currently closer to that activity since I was invited to chair the National Ignition Campaign, Technical Review Committee (NIC-TRC) about 2 years ago. The ICF program was originally established as a means of possibly creating conditions that cause a controlled thermonuclear fusion release by using lasers to generate X-rays that implode a small (~1 mm) capsule. This approach has been pursued more for the possibility that it would produce information to that would be useful in understanding the physics of nuclear weapons as part of our Stockpile Stewardship Program (SSP).

Since its commissioning in May of 2009, the National Ignition Facility (NIF) has been utilized to make realistic progress toward the challenging goal of ICF. I am not aware of any physics reason why this approach can't work, but it is a difficult task. This NIC program is currently operating under conditions that have a goal of certain accomplishments that need to be completed in fiscal year 2012. In addition to its goal of ignition, ICF is also a candidate for possible generation of energy for generating electricity.

Much as with magnetic fusion, generating electricity using the ICF approach also needs to be able to release more energy than that required for ignition if it is to be commercially viable. Both approaches to fusion should be pursued at a level that allows a demonstration of what is possible before committing to a particular approach to electric power generation or some other process involving the production of bred fissile atoms from fusion neutrons.

There are some who say we don't know enough of the basic science and technology to proceed and that we should focus on them until we have a better understanding of all of the fundamental processes.

Suppose that when someone noticed that we could make steam with fire, it was insisted by some that we should understand the physics of flame before proceeding to make a steam locomotive. That would have possibly stopped the development of steam engines. It has only been in the last couple of decades that we have come to understand the physics of flames using lasers as a diagnostic tool. We only needed the knowledge in Keenan and Keeves' "Steam Tables" and how to make a boiler that didn't blow up to make a major industry out of steam engines.

For fusion, we worry about understanding certain theoretical models to examine the physics of plasmas. This is certainly a worthy goal. Even so, I doubt that when a working fusion reactor is built that there will be a detail understanding of processes that can be explained by quasi-linear perturbation theory or magnetohydrodynamics. Rather, it is likely that some opaque process involving large amplitude turbulence will lie at the heart of the reactor, and we may never know why it works as well as it does. That is, it might be like the use of fire.

Alvin W. Trivelpiece has played many roles in his career that have both directly and indirectly influenced the course of fusion research, including coauthor (with Nicholas Krall) of a graduate level textbook on plasma physics, associate director for Research in the AEC CTR division, director of the DOE Office of Energy Research (now Office of Science), and director of Oak Ridge National Laboratory. He has also served as a consultant to several national laboratories since his retirement.

The Ultimate Energy Source? 15

Think you can or think you can't. Either way you will be right.

–Henry Ford

If you can dream it, you can do it.

–Walt Disney

Stronger than all the armies on earth is an idea whose time has come.

–Senator Everett Dirksen

Politics

Currently, governments fund most fusion research. In the USA, progress has been severely constrained by the availability of federal research dollars and an absence of policy decisions to construct new fusion facilities. For example, following the success of the US TFTR in the mid-1990s, Congress cut fusion funding rather than to authorize the construction of a more advanced facility. Congress, at the time, had embarked on a "cut federal spending in general" budget frenzy that periodically dominates Washington thinking. As this book goes to press (September 2012), we are in the midst of another such frenzy, brought about by a desire "to cut the federal deficit."

The US Department of Energy has to balance priorities among many competing programs. These include such disparate areas as weapons, energy technologies, energy efficiency, and basic science. The priorities change as new administrations take office, and these priorities often are based more on political agendas than on logic or national need. The same can be said for how the Office of Management and Budget (OMB) sets priorities among government agencies. As an arm of the White

S.O. Dean, *Search for the Ultimate Energy Source: A History of the U.S. Fusion Energy Program*, Green Energy and Technology, DOI 10.1007/978-1-4614-6037-4_15,

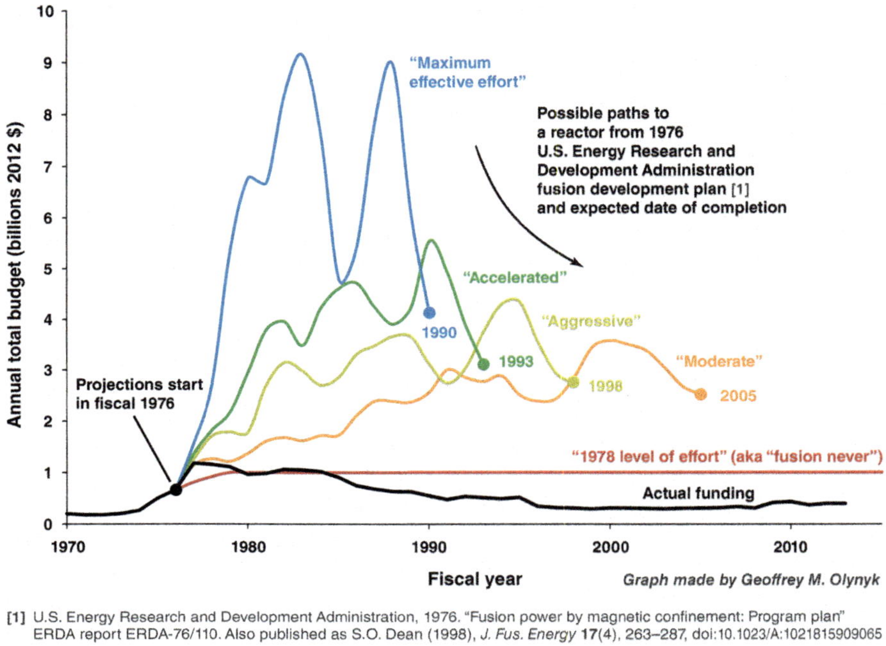

Fig. 15.1 US fusion budget history compared to five plans proposed in 1976. Since 1980, neither the funding nor the construction of facilities needed to move the USA toward a fusion power plant has been forthcoming

House, the OMB is even more prone to priority shifts in order to fulfill campaign promises of the current president.

Much time and expense is devoted in government to asking for advice from advisory committees. These committees usually do a good, objective job. Unfortunately, their advice is rarely followed since the advice frequently flies in the face of political agendas or simply is viewed as costing too much money.

The politics of changing US federal energy policies has regularly stymied all attempts by the fusion community to capitalize on fusion scientific successes over the past several decades. As shown in Fig. 15.1, the US fusion budget remains on a track that does not lead to a US demonstration fusion power plant ever. As the US budget has stagnated, however, the world fusion effort has mushroomed. During the 1980s, the European and Japanese fusion programs approximately doubled and, in more recent years, the fusion programs in Korea and China have burgeoned. High-level government commitments in the latter countries have been made to fusion energy development [149, 150].

Figure 15.1 traces the history of the civilian (magnetic) fusion energy program. The inertial fusion program is funded in the DOE weapons budget category and has largely been exempt from budget fluctuations. Weapons budget managers have regularly refused to request funding for developing the technologies required for civilian energy applications of inertial fusion. However, Congress did provide, and the DOE weapons program did implement, a civilian energy-oriented high-average power laser (HAPL) program from the late 1990s to the late 2000s. That program has since been terminated. Currently, DOE has asked the National Academies of Science and Engineering for recommendations on what kind of inertial fusion energy development program should be initiated if and when NIF lasers ignite a fusion capsule. Whether the required funding will then be available to implement a program to develop inertial fusion for energy in the current budget-cutting atmosphere in Washington remains to be seen.

Progress

Over the past approximately 60 years, steady progress, consistent with funding provided, has been achieved in producing fusion energy in a controlled manner in both magnetic and inertial fusion laboratory experiments. The progress toward the fusion goal is shown in Fig. 15.2. The primary methods have been the use of magnets to form tokamak magnetic bottles and using lasers to compress capsules containing fusion fuel.

Over 10 MW of fusion power was produced for a few seconds in the mid-1990s in the Tokamak Fusion Test Reactor (TFTR) in the USA and in the Joint European Torus (JET) in the UK. Another large tokamak facility, the JT-60 in Japan, achieved similar

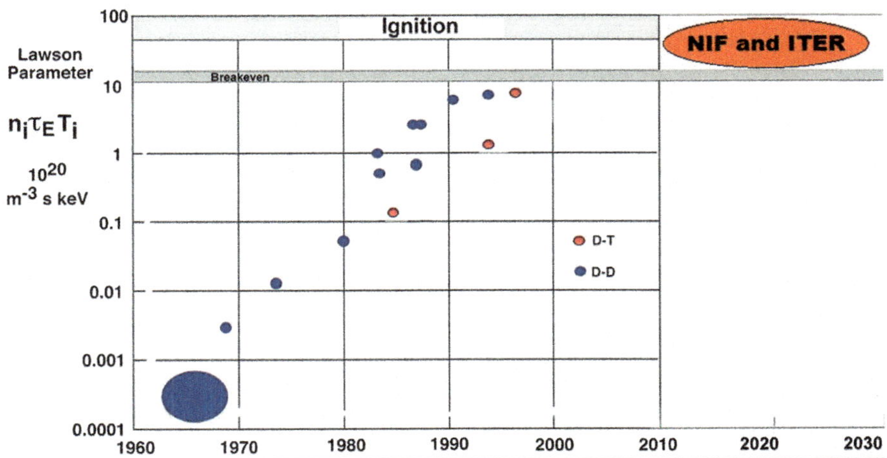

Fig. 15.2 Progress in achieving the Lawson "triple product" parameter required for fusion power plants in both magnetic and inertial fusion laboratory experiments is shown on the graph for both deuterium-only (DD) and deuterium–tritium (DT) fuel mixtures. NIF began operation in 2009 and hopes to achieve a result at or above the breakeven line; ITER, which was at one time to operate by the year 2000, is now expected to begin operation around 2020, although operation using deuterium–tritium (DT) fuel is not expected until around 2027

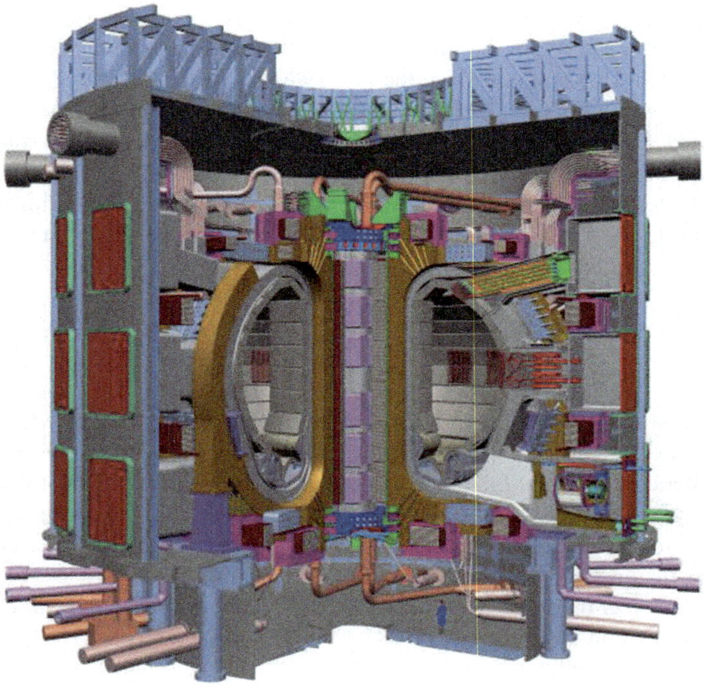

Fig. 15.3 ITER Drawing. *Note* nominal six-foot person, *lower right* (ITER Organization: http://www.iter.org)

plasma conditions in deuterium (DD). The next major step on the road to tokamak fusion power is projected to occur in the International Thermonuclear Experimental Reactor (ITER) now under construction in France. An artist's drawing of ITER is shown in Fig. 15.3. ITER is a joint venture among the European Union, China, India, Japan, Korea, Russia, and the USA. Its international character slowed decision making in its early planning and is also now complicating its construction.

Inertial fusion is the other primary approach to fusion. It is being pursued in several countries, most notably the USA, Japan, and the EU, using lasers (or other intense sources of pulsed power) to compress and heat small capsules containing fusion fuel. These efforts began in the early 1960s and have also made steady progress. Currently, a major demonstration of "igniting" a capsule is anticipated at the National Ignition Facility (NIF), located at the Lawrence Livermore National Laboratory in California, shown in Fig. 15.4. The 192-beam NIF laser is the largest high-power laser ever built. A similar facility is under construction in France.

Even after the NIF and ITER facilities have advanced the fusion quest, there will still be much engineering and technology development required before a commercial fusion power plant can operate. ITER, for example, is designed to operate at the 500 MW level but only for 15 min at a time. Technologies are under development to

Fig. 15.4 Photo showing a portion of NIF's 192 laser beams (Lawrence Livermore National Laboratory, https://lasers.llnl.gov/multmedia/photo_gallery/)

allow for extension to steady state, however, and it is hoped by some that the follow-on facility to ITER could be a commercial power plant prototype. NIF is expected to release 5–10 times more fusion energy than the energy input from the laser on a single shot. For a power plant, one will require a laser (or other "driver") capable of igniting capsules several times per second, each releasing about 100 times more fusion energy than the energy input. Technologies and facilities to do just that, and to address the other engineering/technologies needed for commercial fusion, are on the drawing boards but are not yet funded.

Promise

Fusion is a nuclear process that releases about ten million times more energy per pound of matter consumed than in a typical chemical process like the burning of fossil fuels and about eight times more energy per pound of matter consumed than nuclear fission. Fusion is not easily achieved, however. In the Sun and other stars, gravitational forces produce the high densities and temperatures required for the fusion reactions to occur, but on earth, we must work harder to accomplish it.

Fusion has sometimes been called "the ultimate energy source." This is because fusion, the combining of the nuclei of hydrogen isotopes, forms helium (a harmless and valuable "waste" product) and because the deuterium hydrogen isotope can be obtained inexpensively from water and hence will be readily available to all nations for as long as there is life on the planet.

The first large-scale release of fusion energy on earth was in the form of the hydrogen bomb in the early 1950s. Since that time, scientists around the world have sought means to produce fusion energy in a controlled manner for the production of electricity and for other possible applications like desalination of water.

My opinion that fusion will ultimately become a practical source of energy is based on my belief that fusion will ultimately be required to fulfill society's energy needs and my confidence that fusion is technically feasible. Although other energy sources (like nuclear fission and "renewables") might be able to fulfill those needs, their roles are not guaranteed. Public acceptance of fission and fossil sources is uncertain, due to concerns about radioactive waste disposal and nuclear weapons proliferation in the case of fission, to concerns about climate change due to carbon emissions in the case of fossil fuels, and to uncertain economics and/or practicality of providing forecasted large energy demand in the case of solar, wind, and other renewables.

Although not without problems of its own (see Chap. 12), fusion has many potential advantages vis-à-vis other energy options. The fuel supply is indigenous to all nations and is essentially inexhaustible, and fusion has low environmental impact compared to continued reliance on fossil fuels. While nuclear fission is a proven source of electricity, fusion is believed to have superior safety, nuclear proliferation, and radioactive waste characteristics, and compared to solar, wind, and other non-hydro renewables, fusion is more compatible with the large-scale energy needs of an expanding industrial society.

Estimates of when fusion will be ready for commercialization vary widely (from 15 to 50 years), depending on the optimism of particular advocates and assumptions on how much money will be available for development. When I began my own career in fusion in 1962, I believed that I would see fusion power on the electric grid by now. I now think it more likely to be in the 2040–2050 time frame, and more likely first to be in Europe or Asia rather than in the USA.

It is fortunate that the world is not likely to run out of energy sources by then. Issues associated with the cost of energy and climate change will likely become more and more important as time goes on. At some point in the future, fusion will enter the energy mix, at which point we will be able to truly assess its right to be called "the ultimate energy source."

Epilogue

Those who come after us will fill up the canvas we have begun.
–Thomas Jefferson

The fusion story presented here is largely one from my personal perspective and experience. The broader story involves the contributions and perspectives of thousands of dedicated scientists and engineers worldwide. Their contributions are documented in other books and in countless journal articles and scientific/engineering conference proceedings. I have described in this book the contributions and roles of relatively few. Many are still active, some are retired, and some, sadly, have passed away.

When I left government in 1979 after 17 years of service, to form Fusion Power Associates (FPA), it was my hope that the US government would begin the transition from a science-dominated program to one that included engineering development in industry. With the passage of the Magnetic Fusion Energy Engineering Act of 1980 [34], many industries and electric utilities rallied to the fusion cause by joining Fusion Power Associates, as shown in Tables 4.3 and 5.4. Unfortunately, as the budget profile shown in Fig. 15.1 indicates, the US government did not rise to the occasion then or subsequently. The US fusion research community steadfastly advocated keeping an eye on the energy goal, however, as illustrated by their steadfast support of the goals of Fusion Power Associates. Lists of the current institutional members and board of directors of FPA are shown in Tables 9.1 and E.1.

Fusion Power Associates began recognizing many of the "leaders" of the fusion effort, and those who have made special contributions, by establishing leadership, distinguished career, excellence in fusion engineering, and special awards. Leadership awards were established in 1980 "to recognize individuals who have provided outstanding leadership in accelerating the development of fusion power." Distinguished Career awards were established in 1987 "to recognize individuals who have made distinguished, lifelong career contributions to fusion development." Excellence in Fusion Engineering awards were established in 1987, in memory of MIT Professor

S.O. Dean, *Search for the Ultimate Energy Source: A History of the U.S. Fusion Energy Program*, Green Energy and Technology, DOI 10.1007/978-1-4614-6037-4,

Table E.1 Fusion Power Associates Board of Directors and Officers: 2012

Farokkh Najmabadi, University of California, San Diego, Chair
B. Grant Logan, Lawrence Berkeley National Laboratory, Vice Chair
Mohamed Abdou, University of California, Los Angeles
David Babineau, Savannah River National Laboratory
E. Michael Campbell, Logos Technologies
Donald L. Correll, Lawrence Livermore National Laboratory
N. Anne Davies, US Department of Energy (retired)
Gerald L. Kulcinski, University of Wisconsin
M. Keith Matzen, Sandia National Laboratories
Robert L. McCrory, University of Rochester
Dale M. Meade, Princeton Plasma Physics Laboratory (retired)
Stanley Milora, Oak Ridge National Laboratory
Edward Moses, Lawrence Livermore National Laboratory
Gerald Navratil, Columbia University
Stephen P. Obenschain, US Naval Research Laboratory
Nicole Petta, Schafer Corporation
Miklos Porkolab, Massachusetts Institute of Technology
Stewart Prager, Princeton Plasma Physics Laboratory
John Sheffield, University of Tennesee (retired)
Tony S. Taylor, General Atomics
Francois Waelbroeck, University of Texas, Austin
Thomas Weaver, The Boeing Company
Glen Wurden, Los Alamos National Laboratory
Officers:
Stephen O. Dean, President
Ruth A. Watkins, Vice President, Administration and Finance
Mark S. Tillack, Vice President, Communications

David J. Rose, "to recognize persons in the relatively early part of their careers who have shown both technical accomplishment and potential to become exceptionally influential leaders in the fusion field." Special awards have been given periodically to persons "who have made important contributions to the fusion effort." The recipients of these awards are listed in Tables E.2, E.3, E.4, and E.5.

Despite steady progress in fusion scientific understanding and experimental demonstration over the past 60 years, commercial success is not guaranteed. In the future, much more attention will have to be given to developing fusion plant embodiments that meet commercial competitiveness standards of the marketplace for the specific applications discussed in Chap. 11. These will be affected by government regulations, which will certainly vary in different countries.

This book went to press in early September 2012. For an update on events since then, visit Fusion Power Associates web site: http://fusionpower.org and click on Fusion Program Notes; also see postings on the FIRE site: http://fire.pppl.gov.

Table E.2 FPA Leadership Award Recipients

1980	**1992**	**2003**
Solomon J. Buchsbaum	Robert Conn	Stewart Prager
Robert L. Hirsch	Gerald Kulcinski	**2004**
Mike McCormack	**1993**	Raymond Fonck
Paul Tsongas	Donald Cook	Farrokh Najmabadi
1981	John Sheffield	**2005**
Edwin E. Kintner	**1994**	Ronald Stambaugh
1982	Charles Baker	**2006**
Harold P. Furth	Steven Koonin	Gerald Navratil
John H. Nuckolls	**1995**	Ned Sauthoff
1983	E. Michael Campbell	**2007**
John Emmett	David Overskei	Richard Hawryluk
T. Kenneth Fowler	**1996**	**2008**
1984	Mohamed Abdou	Edward Moses
Tihiro Ohkawa	Robert McCrory	Tony Taylor
Gerold Yonas	**1997**	**2009**
1985	David E. Baldwin	Gyung-Su Lee
Evgenii P. Velikhov	**1998**	**2010**
Chiyo Yamanaka	Hermann Grunder	Riccardo Betti
1986	John P. Holdren	Y-K Martin Peng
Ronald C. Davidson	**1999**	**2011**
1987	B. Grant Logan	M. Keith Matzen
Marshall N. Rosenbluth	Dale Meade	**2012**
1988	**2000**	Stephen P. Obenschain
John Clarke	Robert Aymar	
1989	John Lindl	
Paul-Henri Rebut	**2001**	
1990	Robert J. Goldston	
Boris B. Kadomtsev	Ronald R. Parker	
1991	**2002**	
Bruno Coppi	Richard Hazeltine	
Erik Storm	Jeffrey Freidberg	
	John Sethian	

Established in 1980; presented to individuals who have shown outstanding leadership qualities in accelerating the development of fusion

Table E.3 FPA Distinguished Career Award Recipients

1987	**1995**	**2004**
Melvin B. Gottlieb	T. Kenneth Fowler	Bruno Coppi
Donald Kerst	Harold P. Furth	**2005**
Richard F. Post	**1996**	Charles Baker
Lyman Spitzer, Jr.	Joseph G. Gavin	Dale Meade
1988	John H. Nuckolls	**2006**
K. Husimi	**1997**	N. Anne Davies
Donato Palumbo	Marshall N. Rosenbluth	Vladimir Tolok
R. Sebastian Pease	**1998**	**2007**
1989	Boris B. Kadomtsev	David E. Baldwin
Frederick H. Coensgen	D. Bruce Montgomery	**2008**
Donald J. Grove	Tihiro Ohkawa	Osamu Motjima
Fred L. Ribe	Paul Rutherford	**2009**
1990	**1999**	Weston M. Stacey, Jr.
Nicolai G. Basov	Thomas H. Stix	**2010**
Tedashi Sekiguchi	J. Bryan Taylor	Miklos Porkolab
1991	Masaji Yoshikawa	Dmitri Ryutov
Harold K Forsen	**2000**	**2011**
John W. Landis	Alan Gibson	Ronald R. Parker
Robert L. Sproull;	Thomas Simonen	**2012**
A.Guyford Stever	Ken Tomabechi	B. Grant Logan
1992	Alvin W. Trivelpiece	
Roy Bickerton	**2001**	
Amasa S. Bishop	Roger O. Bangerter	
V. A. Glukikh	Edward A. Frieman	
Sigeru Mori	**2002**	
1993	James D. Callen	
Robert A. Gross	Gunter Grieger	
Murray W. Rosenthal	Sadao Nakai	
1994	**2003**	
Charles A. Flanagan	Robert Aymar	
Wulf G. Kunkel	John Sheffield	

Established in 1987; presented to individuals who have made distinguished, lifelong, career contributions to fusion development

Table E.4 FPA Excellence in Fusion Engineering Award Recipients

S. J. Piet	F. Najmabadi	A. Nikroo
M. A. Ulrichson	G.G. Denisov	C. Bibeau
D. Ehst	P.J. Gierjewski	N. Morley
Y-K Martin Peng	P. Barabaschi	Y. Katoh
W. Reirson	S. Payne	B. Wirth
J. Santarius	M. Tillack	J.P. Sharpe
O. Filatov	P.F. Peterson	D. Garnier
S. Zinkle	M.D. Williams	J. Latkowski
J.D. Galambos	G. Federici	P. Patel
S. W. Haney	M. Mauel	M. Dunne
C. E. Kessel	L. L. Snead	M. Herrmann
K. A. McCarthy	R. Kodama	

Established in 1987 in memory of MIT professor David J. Rose; presented to individuals in the early part of their careers who have shown both technical accomplishment and potential for becoming exceptionally influential leaders in the fusion field

Table E.5 FPA Special Awards

J. Robert Beyster	Dean Gallo	Paul Rivenburg
Edward A. Frieman	Kathryn M. Thorpe	Paul Thomas
Henry J. Gomberg	Marilyn Lloyd	Nan Wells
Bernard J. Eastlund	Marshall Sluyter	John DeLooper
Sherman Naymark	Roscoe Bartlett	Michael Roberts
Glenn Sorenson	Rodney Freylinghuysen	Tihiro Ohkawa
Alvin W. Trivelpiece	Tim Roemer	John H. Nuckolls
Paul J. Reardon	Rush Holt	Richard F. Post
James M. Williams	Ron Packard	Jeff Hoy
Gerold Yonas	Diane Carroll	Darlene Markevich
Donald P. Zeifang	Donald Correll	Christopher J. Keane
Ruth Ann Watkins	Carol Danielson	John W. Willis
George S. Clemens	Stephen O. Dean	Erol Oktay
John Killeen	Mark Haynes	

Established in 1980; presented periodically to recognize individuals who have made some special contribuion(s) to the cause of fusion power development

Acknowledgements

I express my admiration, thanks, and gratitude to the fusion scientists and engineers with whom I have had the pleasure of working over the years, many of whom are mentioned in this book. I particularly thank those who generously contributed their own current perspectives in Chap. 14.

I especially express my appreciation to Ms. Ruth Ann Watkins, who was my secretary at the US Atomic Energy Commission many years ago, joined me to help form Fusion Power Associates in 1979, and has helped keep the fusion vision alive to the present day as FPA vice president for administration and finance.

Recommended Reading and Information Resources

Bishop A.S.: Project Sherwood. Addison Wesley Publishing Company, Boston, MA (1958)

Breizman B.N., Van Dam J.W., editors: G. I. Budker – reflections & remembrances. AIP Press, New York (1994)

Bromberg J.L.: Fusion—science, politics, and the invention of a new energy source. MIT, Cambridge, MA (1982)

Chen F.F.: An Indispensable Truth—how fusion power can save the planet. Springer, Heidelberg (2011)

DOE Office of Fusion Energy Sciences Web Site: http://science.energy.gov/fes

Fusion Power Associates Web Site: http://fusionpower.org which also includes links to other fusion sites worldwide

FIRE Web Site: http://fire.pppl.gov

Glasstone S., Lovberg R.H.: Controlled Thermonuclear Reactions. D. Van Nostrand Company, Princeton, NJ (1960)

Heppenheimer T.A.: The Man-Made Sun. Little, Brown and Company, New York (1984)

Herman R.: Fusion—The search for endless energy. Cambridge University Press, Cambridge (1990)

Hirsch R.L., Bezdek R.H., Wendling R.M.: The Impending World Energy Mess: what it is and what it means to you! Apogee Prime Press (2010)

Kaku M., Trainer J., editors: Nuclear power: Both Sides. W. W. Norton & Company, New York (1982)

Krivit S.B., editor, Wiley series on energy, Nuclear Energy Encyclopedia. Wiley, New York (2011)

McCracken G., Stott P.: Fusion—the energy of the universe. Elsevier Academic, New York (2005)

Stacey W.M.: The Quest for a Fusion Energy Reactor—An Insider's Account of the Intor Workshop. Oxford University Press, Oxford (2010)

Velarde G., Santamaria N.C., editors, Inertial Confinement Nuclear Fusion – a historical approach by its pioneers. Foxwell & Davies, UK (2007)

References

1. Bishop, A.S.: Project Sherwood. Addison Wesley Publishing Company, Boston, MA (1958)
2. Glasstone, S., Lovberg, R.H.: Controlled Thermonuclear Reactions—An Introduction to Theory and Experiment. D. Van Nostrand Company, Princeton, NJ (1960)
3. McCracken, G., Stott, P.: Fusion—The Energy of the Universe. Academic, New York (2005)
4. Bromberg, J.L.: Fusion—Science, Politics, and the Invention of a New Energy Source. MIT, Cambridge, MA (1982)
5. Obituary of Oleg Lavrentiev, written by Vladimir Voitsenya, published in Fusion Power Associates Fusion Program Note FPN11-15, 24 March 2011 and posted at http://aries.ucsd.edu/fpa/fpn11-15.stml. See also Lavrentiev bio in Wikipedia
6. Lawson, J.D.: Some Criteria for a Power-producing Thermonuclear Reactor. Proc. Phys. Soc. (London) **B70**(6) (1957)
7. Post, R.F.: Controlled fusion research—an application of the physics of high temperature plasmas. Rev. Mod. Phys. **28**, 338–362 (1956)
8. Ryutov, D.D., Berk, H.L., Cohen, B.I., Molvik, A.W., Simonen, T.C.: Magneto-hydrodynamically Stable Axisymmetric Mirrors. Physics of Plasmas 18, published online 1 September 2011
9. Simonen, T.C.: Fusion Sci. Technol. **57**, 305 (2010)
10. Velarde, G., Santamaria, N.C. (eds.): Inertial Confinement Nuclear Fusion—A Historical Approach by Its Pioneers. Foxwell & Davies, UK (2007)
11. Freidberg, J.P.: Plasma Physics and Fusion Energy. Cambridge University Press, Cambridge (2007)
12. Bellan, P.M.: Spheromaks. Imperial College Press, London (2000)
13. Rostoker, N., Binderbauer, M.W., Monkhurst, H.J.: Colliding beam fusion reactor. Science **278**, 1419–1422 (1997)
14. Binderbauer, M.W., et al.: Dynamic formulation of a hot field reversed configuration with improved confinement by supersonic merging of two colliding high-beta compact toroids. Phys. Rev. Lett. **105** (23 July 2010)
15. Coleman, E.S., Cohen, S.A., Mahoney, M.S.: Greek fire: Nicholas Christofilos and the Astron Project in America's early fusion program. J. Fusion Energ. **30**, 238–256 (2011)
16. Bishop, A.S., Dean, S.O., Post, R.F.: AEC Policy and Action Paper on Controlled Thermonuclear Research. AEC Report TID-23277 (June 1966); posted at http://fire.pppl.gov/US_AEC_Fusion_Policy_1966.pdf and later published in J. Fusion Energ. **30**, 207–237 (2011)
17. Forrest, M.: Lasers Across the Cherry Orchards. Tandem, Madison, WI (2011)
18. Gough, W.C., Eastlund, B.J.: The fusion torch. Bull. Am. Phys. Soc. **11**, 1564 (1968) and AEC Report WASH-1132 (15 May 1969)

S.O. Dean, *Search for the Ultimate Energy Source: A History of the U.S. Fusion Energy Program*, Green Energy and Technology, DOI 10.1007/978-1-4614-6037-4,

19. Dean, S.O., McLean, E.A., Stamper, J.A., Griem, H.R.: Demonstration of collisionless interactions between interstreaming ions in a laser-produced plasma experiment. Phys. Rev. Lett. **27**, 487 (1971)
20. Dean, S.O., McLean, E.A., Stamper, J.A., Griem, H.R.: Reasons for the collisionless nature of interactions in a laser-produced plasma experiment. Phys. Rev. Lett. **29**, 569 (1972)
21. Stamper, J.A., Papadopoulos, K., Sudan, R.N., Dean, S.O., McLean, E.A.: Spontaneous magnetic fields in laser-produced plasma. Phys. Rev. Lett. **26**, 1012 (1971)
22. Dean, S.O.: Confinement of Laser-Produced Plasma in Resonant Cavities by RF Electromagnetic Field. NRL Report 7136 (28 September 1970)
23. Dean, S.O.: Laser-Generated Fusion Plasmas. Report of NRL Progress (December 1971)
24. Dean, S.O.: Fusion Power by Magnetic Confinement: Program Plan. U.S. Energy Research and Development Administration Report ERDA-76/110 (July 1976) posted at http://fire.pppl.gov/us_fusion_plan_1976.pdf. Also published in J. Fusion Energ. **17**, 263 (1998)
25. Krall, N.A., Trivelpiece, A.W.: Principles of Plasma Physics. McGraw-Hill Book Company, New York (1973)
26. Dean, S.O., Callen, J. D., Furth, H. P., Clarke, J. F., Ohkawa, T., Rutherford, P. H.: Status and Objectives of Tokamak Systems for Fusion Research. U.S. Atomic Energy Commission Report WASH-1295 (1974) posted at http://fire.pppl.gov/wash_1295.pdf. Also published in J. Fusion Energ. **17**, 289 (1998)
27. Nuckolls, J., Wood, L., Thiessen, A., Zimmerman, G.: Laser compression of matter to super-high densities: CTR applications. Nature **239**, 139 (1972)
28. Dean, S.O. (Mirror Senior Review Panel, Chairman), Bussard, R., Fraas, A., Forsen, H., Miller, B., Rosenbluth, M., Trivelpiece, A., Williams, J., Grace, J.: An Assessment of the Role of Magnetic Mirror Devices in Fusion Power Development. ERDA 76–64 (April 1976), including, as an Appendix, approval statement by the Fusion Power Coordinating Committee, E. E. Kintner, Chairman (21 April 1976)
29. Kenneth Fowler, T.: The Fusion Quest. The Johns Hopkins University Press, Baltimore, MD (1997)
30. Stacey, W.M.: The Quest for a Fusion Energy Reactor. Oxford University Press, Oxford (2010)
31. U. S. Makes Major Advance in Nuclear Fusion, Banner Headline and Front Page Story in the Washington Post (Sunday, 13 August 1978)
32. Heppenheimer, T.A.: The Man-Made Sun. Little, Brown and Company, New York (1984)
33. Freeman, M., Jones, W.: Fusion in Korea: energy for the next generation. Exec. Intell. Rev. **36** (4 December 2009)
34. Magnetic Fusion Energy Engineering Act of 1980, Public Law 96–386 (7 October 1980) published in J. Fusion Energ. **1**, 149 (1981). Also posted at http://fire.pppl.gov/mfe_act_1980.pdf
35. Dean, S.O. (ed.): Prospects for Fusion Power. Pergamon Press, Oxford (1981)
36. Heppenheimer, T.A.: The Man-Made Sun. Little Brown and Company, New York (1984)
37. Kintner, E.E.: Casting fusion adrift. MIT Technol. Rev. (May/June 1982)
38. Utility Requirements for Fusion. EPRI Topical Report AP-2254 (February 1982)
39. An Assessment of the U. S. Mirror Fusion Program: Report of the 1980 Mirror Senior Review Panel. DOE/ER-0057 (February 1980)
40. Reports of the Magnetic Fusion Advisory Committee. J. Fusion Energ. **7**, 227 (1988)
41. Dean, S.O.: Fusion 1983: A symposium on the readiness and reasons for an Accelerated National Development Program—summary report. J. Fusion Energ. **3**, 151 (1983)
42. Greenwald, M., et al.: Energy confinement of high-density pellet-fueled plasmas in the Alcator-C Tokamak. Phys. Rev. Lett. (23 July 1984), also reported in the Philadelphia Inquirer (9 November 1983)
43. Goldston, R.J.: Plasma Phys. Control. Fusion **26**, 87 (1984)
44. ITER physics basis. Nuclear Fusion **36**, 2137 (1999)
45. Magnetic Fusion Energy R&D: A Report of the Energy Research Advisory Board to the United States Department of Energy (February 1984)

46. Fusion Power Associates Executive Newsletter (August 1984)
47. Where Do We Go from Here? Interview with Trivelpiece A.W. Fusion Power Associates Executive Newsletter (September 1984) posted in the Library Section at http://fire.pppl.gov/trivelpiece_1984_interview.pdf
48. Fusion Power Associates Executive Newsletter (August 1985)
49. Technical Planning Activity Final Report. Argonne National Laboratory ANL/FPP-87-1 (January 1987)
50. Dean, S.O.: Commercial objectives, technology transfer and systems analysis for fusion power development. J. Fusion Energ. **17**, 25 (1988)
51. Fusion Power Associates Executive Newsletter (December 1985)
52. Fusion Power Associates Executive Newsletter (February 1986)
53. Fusion Power Associates Executive Newsletter (March 1986); also see articles by Mark Crawford in Science (23 May 1986 and 7 Nov 1986)
54. Fusion Power Associates Executive Newsletter (August 1985)
55. Fusion Power Associates Executive Newsletter (September 1985)
56. Fusion Power Associates Executive Newsletter (June 1986)
57. Fusion Power Associates Executive Newsletter (May 1988)
58. Fusion Power Associates Executive Newsletter (January 1989)
59. Dean, S.O.: Status of Candidate Drivers for a Laboratory Microfusion Facility (LMF). Final Report to the U.S. Department of Energy (April 1988; updated July 1989); summary published in Fusion Technology (American Nuclear Society) (March 1989)
60. Inertial Confinement Fusion Program Plan Summary for Fiscal Years 1990–1994. DOE Report DOE/DP/IFD/PP090189 (1989)
61. Fusion Power Associates Executive Newsletter (December 1989)
62. Fusion Power Associates Executive Newsletter (February 1990)
63. Hora, R.P., Locke Bogart, S.: J. Fusion Energ. **11**, 225 (1992)
64. Manifold Productions, Fire from the Sun, PBS film available on DVD from http://www.manifoldproductions.com/FireSunFilm.html
65. Fusion Policy Advisory Committee (FPAC) Final Report, USDOE Report DOE/S-0081 (September 1990); also posted in the library section at http://fire.pppl.gov/fpac_1990.pdf and also published in J. Fusion Energ. **10**, 127 (1991)
66. Fusion Power Associates Executive Newsletter (October 1990)
67. Fusion Power Associates Executive Newsletter (July 1992)
68. Dean, S.O., Baker, C.C., Cohn, D.R., Kinkead, S.D.: An accelerated fusion power development plan. J. Fusion Energ. **10**, 197 (1991)
69. Galambos, J., Baker, C., Peng, Y.-K.M., Cohn, D., Chaniotakis, M., Bromberg, L., Dean, S.: Systems studies of copper- and superconducting-coil pilot plants. Fusion Technol. **21**(1759) (1992)
70. Dean, S.O., Baker, C.C., Cohn, D.R., Dautovich, D.P., Ellis, W.R., Morison, W.G.: Pilot plant: an affordable step toward fusion power. J. Fusion Energ. **11**, 99 (1992)
71. Dean, S.O., Baker, C.C., Galambos, J., Martin Peng, Y-K., Sheffield, J., Cohn, D., Bromberg, L., Chaniotakis, E.A., Dautovich, D.P., Gierszewski, P., Morison, W., Ellis, W.R., Kadomtsev, B.: Pilot plant: A Shortened Path to Fusion Power. Proceedings of Fourteenth International Conference on Plasma Physics and Controlled Nuclear Fusion Research, Paper IAEA-CN-56/G-1-5 (International Atomic Energy Agency, Vienna, Austria)
72. J. Fusion Energ. **11** (December 1992)
73. Fusion Power Associates Executive Newsletter (September 1991)
74. Fusion Power Associates Executive Newsletter (December 1991)
75. Paul Gilman, quoted in Los Alamos National Laboratory Newsbulletin (22 February 1991)
76. Hirsch, R.L., Culler, F., Hingorani, N.G., Taylor, J.J., Schneider, T.R., Spencer, D.R., Report of the 1992 EPRI Fusion Review Panel. EPRI TR-1-1649 (November 1992); published in J. Fusion Energ. **11**, 209 (1992)

77. Kaslow, J., et al.: Criteria for Practical Fusion Power Systems; Report from the EPRI Fusion Panel. EPRI Report BR-104469 (Spring 1994); also published in J. Fusion Energ. **13**, 181 (1994)
78. Fusion Power Associates Executive Newsletter (January 1995)
79. Holdren, J.P., et al.: An Assessment of Environmental, Safety, and Economic Aspects of Fusion. LLNL Report UCRL-53766 (1989)
80. Conn, R.W., et al.: A restructured fusion energy sciences program. J. Fusion Energ. **15**(183) (1996)
81. Holdren, J.P., et al.: Report to the President on Federal Energy Research and Development for the Twenty-first Century (November 1997); posted at http://fire.pppl.gov/pcast_1997.pdf
82. Holdren, J.P., et al.: The U.S. Program of Fusion Energy Research and Development: Report of the Fusion Review Panel of the President's Council of Advisors on Science and Technology. J. Fusion Energ. **14**, 213 (1995)
83. Dean, S.O., et al.: Pathways to fusion power. J. Fusion Energ. **17**(1) (1998)
84. Dean, S.O., et al.: Cost-effective steps to fusion power. J. Fusion Energ. **17**(177) (1998)
85. Meserve, R.A., et al.: Realizing the promise of fusion energy: final report of the task force on fusion energy. J. Fusion Energ. **18**, 85 (1999); posted at http://fire.pppl.gov/SEAB_final_Nov99.pdf
86. Dean, S.O.: Fusion science and technology for the new millennium. J. Fusion Energ. **18**, 1 (1999)
87. Statement on Fusion and Energy Policy (5 February 2001). Posted at http://fusionpower.org
88. An Assessment of the Department of Energy's Office of Fusion Energy Sciences Program. National Academies Press (1 January 2001)
89. Dean, S.O.: Fifty Years of U.S. Fusion Research—An Overview of Programs. Nuclear News (American Nuclear Society, July 2002)
90. Goldston, R.J., et al.: A plan for the development of fusion energy (5 March 2003) posted at http://fire.pppl.gov/fesac_dev_path_final_rpt.pdf. Also published in J. Fusion Energ. **21**, 61 (2002)
91. Linford, R., et al.: A review of the U.S. department of energy's inertial fusion energy program. J. Fusion Energ. **22**, 93 (2003)
92. 2007 IFE Workshop Proceedings. Posted at http://ifeworkshop.llnl.gov/proceedings.html
93. Meade, D.: 50 years of fusion research. Nuclear Fusion **50**, 1 (2010) posted at http://fire.pppl.gov/nf_50th_6_Meade.pdf
94. Dean, S.O.: The rationale for an accelerated fusion energy program. J. Fusion Energ. **27**, 149 (2008)
95. Davidson, R.C., Kulcinski, G., et al.: An Assessment of the Prospects for Inertial Fusion Energy: Interim Report of the NAS Committee on the Prospects for Inertial Confinement Fusion Energy Systems. Posted at http://fire.pppl.gov/NAS_ICF_interim_review_2012.pdf
96. Dunne, M., et al.: Timely delivery of laser inertial fusion energy (LIFE). Fusion Sci. Technol. **60**, 19 (2011); Also, Mike Dunne, *LIFE Power Plant*. Paper presented at Fusion Power Associates 31st Annual Meeting and Symposium (14–15 December 2011). Posted at http://fire.pppl.gov/fpa11_Dunne_Life.pdf
97. Dean, S.O., et al.: Report of FEAC panel 3: concept improvement. J. Fusion Energ. **11**(163) (1992)
98. Najmabadi, F., et al.: Report of the subpanel to FESAC concerning alternate concepts. J. Fusion Energ. **18**(161) (1999)
99. Hsu, S.C., et al.: Spherically imploding plasma liners as a standoff driver for magnetoinertial fusion. IEEE Trans. Plasma Sci. **40** (May 2012)
100. Hsu, S.: Plasma Jet Driven Magneto-Inertial Fusion. Fusion Power Associates 32nd Annual Meeting and Symposium (14–15 December 2011). Posted at http://fire.pppl.gov/fpa11_Hsu_MIF.pdf
101. Wurden, G.: Magneto-Inertial Fusion. Fusion Power Associates 32nd Annual Meeting and Symposium (14–15 December 2011). Posted at http://fire.pppl.gov/fpa11_Wurden_MIF.pdf
102. Chang, P.Y., et al.: Phys. Rev. Lett. (15 July 2011)

103. Slutz, S.A., Vesey, R.A.: Phys. Rev. Lett. (13 January 2012)
104. Liner Fusion Workshop (5–8 February 2012). Posted at http://sandia.gov/pulsedpower/Workshop.html
105. Lindemuth, I.R.: Am J. Phys. **77** (May 2009) and paper presented at Fusion Power Associates 31st Annual Meeting and Symposium (1–2 December 2010). Posted at http://fire.pppl.gov/fpa10_fusion_principles_Lindemuth.pdf
106. International Workshop on Magnetic Fusion Energy Roadmapping (2011). Posted at http://advprojects.pppl.gov/roadmapping
107. Chen, F.F.: An Indispensable Truth—How Fusion Power Can Save the Planet. Springer, Heidelberg (2011)
108. McCarthy, K., et al.: Non-electric applications of fusion. J. Fusion Energ. **21**(121) (2002)
109. Freidberg, J.P., Kadak, A.C.: Fusion-fission hybrids revisited. Nat. Phys. **5**, 370 (2009)
110. Martinez-Vale, J.M., et al.: Hybrid Nuclear Reactors. Nuclear Energy Encyclopedia, Chapter 38. Wiley, New York (2011)
111. Moses, E.I., et al.: A sustainable nuclear fuel cycle based on laser inertial fusion energy (LIFE). Fusion Sci. Technol. **56**, 566 (2009)
112. Manheimer, W.: The fusion hybrid as a key to sustainable development. J. Fusion Energ. **23**, 223 (2004)
113. Bethe, H.A.: Nuclear News **21**, 41 (1978)
114. Bethe, H.A.: The fusion hybrid. Phys. Today **32**, 44 (1979)
115. Jameson, R.A.: Accelerator-driven transmutation of nuclear waste and electric power production. J. Fusion Energ. **12**, 379 (1993)
116. Martin Peng, Y.-K., Cheng, E.T.: Magnetic fusion driven transmutation of waste (FTW). J. Fusion Energ. **12**, 381 (1993)
117. Salvatores, M.: Physics features comparison of TRU burners; fusion/fission hybrids. Accelerator-driven systems and low conversion ration fast reactors. Ann. Nucl. Energ. **36**, 1653 (2009)
118. Gough, W.C., Eastlund, B.J.: The Fusion Torch: Closing the Cycle from Use to Reuse. USAEC Report WASH-1132 (15 May 1969); also Energy, Wastes and the Fusion Torch. Unnumbered USAEC Report (27 April 1971); also Marjorie Mazel Hecht, Fusion Torch Can Create New Raw Materials (1 January 2004) posted at http://Larouchepac.com/node/14726
119. Borisov, A.A., et al.: Fusion power plant for water desalination and reuse. Fusion Eng. Design **58–59**, 1109 (2001)
120. Grant Logan, B.: Can We Afford Ocean Desalination to Double the Fresh Water Supply for the Coming 13 Billion People to Live on a Hot Planet? (unpublished)
121. Dean, S.O., Poltoratskaia, N.: Applications of fusion and plasma device technologies. Plasma Devices Operat. **4** (1995)
122. Dean, S.O.: Applications of plasma and fusion research. J. Fusion Energ. **14**, 251 (1995)
123. Dean, S.O.: Plasmas: The Fourth State of Matter. The World and I Magazine (June 1996)
124. Dean, S.O.: The Plasma Touch. The World and I Magazine (June 1996)
125. FESAC Report on Review of the Fusion Materials Research Program. DOE/ER-0727 (July 1998). Posted at http://science.energy.gov/fes/fesac/reports/
126. FESAC Report on Materials Science and Technology Research Opportunities Now and in the ITER Era (February 2012). Posted at http://science.energy.gov/fes/fesac/reports
127. Abdou, M.: Fusion Nuclear Science and Technology (FNST) Challenges and Facilities on the Pathway to Fusion Energy. Presented at Fusion Power Associates 31st Annual Meeting and Symposium (14–15 December 2011). Posted at http://fire.pppl.gov/fpa_annual_meet.html
128. Zinkle, S.: Materials Science Under Extreme Conditions: Comments on Materials Challenges on the Pathway to Magnetic Fusion Energy. Presented at Fusion Power Associates 31st Annual Meeting and Symposium (14–15 December 2011). Posted at http://fire.pppl.gov/fpa_annual_meet.html
129. Anderson, J.L.: Tritium systems: issues and answers. J. Fusion Energ. **4**, 155 (1985)
130. Petti, D., et al.: ARIES-AT safety design and analysis. Fusion Eng. Design **80**, 111 (2006)

131. Cadwallader, L., El-Guebaly, L.: Safety and Environmental Features. Nuclear Energy Encyclopedia, Chapter 36. Wiley, New York (2011)
132. Hancox, R., Butterworth, G.J.: The Management of Fusion Waste. AEA Culham Laboratory Report AEA FUS61, Presented at 16th Symposium on the Technology of Fusion Energy (3–7 September 1990)
133. Holdren, J.P., et al.: Report of the Senior Committee on Environmental, Safety and Economic Aspects of Magnetic Fusion Energy. UCRL-53766 (September 1989)
134. BP Statistical Review of World Energy, June 2012. http://www.bp.com
135. Hirsch, R.L., Bezdek, R.D., Wendling, R.M.: The Impending World Energy Mess. Apogee Prime (2010)
136. U. S. Energy Information Administration. http://www.eia.gov
137. Murray, J.A.: A Walk Through the National Ignition Facility. ICF Quarterly Report, 95–98, Lawrence Livermore National Laboratory UCRL-LR-105821-97-3 (1997)
138. Moses, E.I., Boyd, R.N., Remington, B.A., Keane, C.J., Al-Ayat, R.: The national ignition facility: ushering in a new age for high energy density science. Phys. Plasmas **16**, 041006 (2009)
139. Haan, S.W., et al.: Point design targets, specifications, and requirements for the 2010 ignition campaign on the national ignition facility. Phys. Plasmas **18**, 051001 (2011)
140. National Ignition Campaign Execution Plan, UCRL-AR-213718, NIF-0111975-AA, Rev. 0 (July 2005)
141. National Research Council (U.S.), Interim Report—Status of the Study an "Assessment of the Prospects for Inertial Fusion Energy". National Academies Press, Washington, DC (2012)
142. Wang, W., Jones, T.B., Harding, D.R.: On-chip double emulsion droplet assembly using electrowetting-on-dielectric and dielectrophoresis. Fusion Sci. Technol. **59**, 240 (2011)
143. McCrory, R.L., et al.: Progress in direct-drive inertial confinement fusion research. Phys. Plasmas **15**, 055503 (2008)
144. Goncharov, V.N., et al.: Demonstration of the highest deuterium-tritium areal density using multiple-picket cryogenic designs on OMEGA. Phys. Rev. Lett. **104**, 165001 (2010)
145. Sangster, T.C., et al.: Shock-tuned cryogenic-deuterium-tritium implosion performance on omega. Phys. Plasmas **17**, 056312 (2010)
146. Zhou, C.D., Betti, R.: Hydrodynamic relations for direct-drive fast-ignition and conventional inertial confinement fusion implosions. Phys. Plasmas **14**, 072703 (2007)
147. Direct drive on the national ignition facility. Phys. Plasmas **11**, 2763–2770 (2004)
148. Collins, T.J.B., et al.: A polar-drive–ignition design for the national ignition facility. Phys. Plasmas **19**, 056308 (2012)
149. Lee, G.S.: The Korean Fusion Program. Presented at Fusion Power Associates 30th Annual Meeting and Symposium (2 December 2009); posted at http://fire.pppl.gov/fpa09_KSTAR_gslee.pdf
150. Wan, Y.: Fusion Research in China. Presented at Fusion Power Associates 31st Annual Meeting and Symposium (1 December 2012); posted at http://fire.pppl.gov/fpa10_AISPP_wan.ppt

Author Index

A
Abdou, M., 89, 114, 152, 157
Abraham, S., 150, 154, 155, 158, 161, 183
Adams, A., 61, 74–76
Ahearne, J., 137, 157, 181
Allison, S., 21
Artsimovich, L., 5, 23, 227
Aston, F., 3
Athy, A., 143

B
Baker, C., 54, 88, 89, 111, 157, 161, 164, 181, 211–214
Baldwin, D., 54, 67, 96, 112, 114, 127, 131
Balzhiser, R., 27, 28, 59, 117, 121
Bangerter, R.O., 141, 181
Basov, N.G., 25, 69
Bauser, E., 43
Beal, J. W., 33, 36
Bennett, W.H., 11
Berk, H., 72
Berman, A., 28
Berry, L., 72
Bethe, H., 3
Biggert, J., 164
Bishop, A., 12, 13, 19, 21–24, 28, 45
Bishop, J., 56, 57
Blackman, M., 4
Blanken, R., 35, 36
Bodner, S., 65
Boehlert, S., 154
Bogart, S.L., 32, 36, 106, 131
Bouquard, M., 64, 71
Bowen, W., 57
Branscomb, L., 22
Briggs, R., 95, 134
Brinkman, W.F., 172, 174, 176, 184, 185, 188
Broad, W., 94, 110
Bromberg, L., 23, 32, 52
Brueckner, K., 25
Brumfiel, G., 155
Buchsbaum, S.J., 63, 66, 72, 73, 120
Burnett, S., 34, 35
Bush, G.H.W., 90, 99, 108, 113, 116, 117, 150, 151, 154–156, 158, 160, 165, 166, 183, 187
Bussard, R.W., 16, 31–33, 35, 42

C
Callen, J., 33, 89, 131, 134
Campbell, M., 141
Carter, J., 31, 52, 53, 61–85, 88, 116, 117
Chen, F.F., 17, 194
Cheney, D., 150
Cho, A., 184
Chu, S., 172, 184
Chung, K., 136
Churchill, W., 1–10
Clarke, J., 29, 33, 34, 43, 44, 51, 63, 67, 71, 94, 96, 99, 101
Coensgen, F., 49
Coffman, F.E., 33, 36, 50
Cohn, D., 88
Colgate, S., 20
Conn, B., 54, 96, 112, 113, 119, 129
Conner, J., 131
Cook, D.L., 175
Coppi, B., 21, 39, 80, 88
Cousins, S., 4
Crandall, D. (Dave), 126
Creutz, E., 20, 21
Culler, F., 112, 117

D
David, E., 28
Davidson, K., 104

S.O. Dean, *Search for the Ultimate Energy Source: A History of the U.S. Fusion Energy Program*, Green Energy and Technology, DOI 10.1007/978-1-4614-6037-4,

Davidson, R., 46, 50, 51, 70–73, 88, 96, 112, 115, 121, 129
Davies, N.A., 34–36, 42, 69, 96, 99, 101, 112, 134, 142, 143, 149, 166, 167, 213–214
Dawson, J., 127
De Hoffman, F., 20
De Vries, R., 80
Dean, S.O., 36, 56, 57, 59, 101, 112, 131, 149, 157
Decker, J.F., 32, 37, 50, 67, 95, 100, 113
Deutch, J., 53, 56, 57, 61
Dirksen, E., 233–238
Disney, W., 233–238
Domenici, P.V., 71, 115, 129, 171–192
Drake, J., 129
Dresselhaus, M., 152, 153
Drew, H., 28
Drucker, P., 193–197
Drummond, W., 21, 22
Duncan, C., 62

E
Eastlund, B., 24, 25
Eddington, A., 3
Edwards, J., 64, 65
Einstein, A., 1, 2
Ellis, W.R. (Bill), 35, 36, 53, 68, 69, 75–77, 116, 131, 214–216
Emmett, J., 27, 28, 71, 72
English, S.G., 28, 35
Everson, D., 131

F
Farnsworth, P.T., 16
Fasella, P., 128, 230
Federici, G., 179
Flatow, I., 199–204
Fonck, R., 157, 167
Ford, G., 45, 88, 107
Ford, H., 233–238
Forrest, M., 23
Forsen, H.K, 54, 105, 107, 116, 119, 132
Foster, J.S., 49, 53, 54, 61, 62, 66, 72, 107
Fowler, K., 21, 34, 48, 67, 72, 75, 95
Freeman, M., 57
Freidberg, J., 153, 195
Frieman, E., 61, 63
Fuqua, D., 64
Furth, H., 20, 23, 33, 43, 56, 59, 96, 101

G
Gavin, J., 70, 132–134
George, T.V., 35, 36
Gibbons, J., 127, 132
Gibran, K., 171–192
Gibson, A., 115
Gilbert, C.F., 93
Gilleland, J., 55, 95, 101
Giller, E., 27
Gilman, P., 115, 171–192
Gingrich, N., 84, 125
Glass, A., 101, 116
Godden, R., 125–147
Goldston, R., 77, 157, 159, 175
Gorbachev, M., 108
Gottlieb, M.B., 34, 40, 42, 44, 56, 70, 71, 96, 107
Gough, W.C. (Bill), 20, 196
Gould, R., 28, 31, 62
Grace, J.N., 37, 51, 52
Grad, H., 20
Greenwald, M., 77, 284
Greiger, G., 55
Griem, H., 28, 45
Gross, R.J., 25
Grove, D., 91, 95
Gwinn, D., 77

H
Hall, R., 154
Happer, W., 91–93, 112–114
Harkness, S.D., 131, 134
Hasegawa, A., 105
Hawryluk, R., 121, 139
Hazeltine, R.D., 134, 152, 156, 159, 216–217
Hecht, M.M., 57
Henning, C.D., 33, 36
Heppenheimer, T., 52, 57, 64, 65
Herb, R., 21
Herbert, F., 105–123
Herrington, J., 89, 93
Hingorani, N., 117
Hirsch, R.L. (Bob), 16, 24, 28, 29, 31–36, 40, 42–44, 46, 47, 49–53, 59, 62, 63, 69, 70, 72, 117–119, 206, 217–219
Hodel, D., 72
Holdren, J.P., 112, 127, 137, 171, 184
Holland, M., 146
Hora, R.P., 106
Hunter, D., 95, 98
Hunter, R., 95, 98–101, 103, 104, 113
Huse, R., 28

I
Iacocca, L., 87–104
Ignat, D., 35, 36
Ikeda, K., 176

J
Jennings, J.D., 92
Jobs, S.P., 27–59
Johnson, M., 35, 36, 53
Johnson, T., 101
Johnston, J.B., 121

K
Kadish, A., 35, 36
Kadomtsev, B., 55
Kantrowitz, A.R., 3
Kennel, C., 134, 152
Kerst, D., 21
Keyworth, G., 64, 65, 80
Kinkead, S., 111
Kintner, E.E., 36, 46, 47, 49–51, 53–57, 63–70, 72, 205–209, 215
Knotec, M., 131–133
Kolb, A., 26
Koonin, S., 103, 120, 172–174, 180, 183, 220
Krall, N.A., 16, 21, 31, 58, 59, 232
Kramer, D., 173
Krebs, M., 133, 134, 145, 152
Kulcinski, G. (Jerry), 16–17, 54, 55, 120, 181
Kurchatov, I., 5, 23, 38

L
Landis, J.W., 57, 107, 131, 132
Lang, J., 131
LaRouche, L., 57
Lavrentiev, O., 5
Lawson, J.D., 7, 9, 34, 91, 93
Leontovich, M., 5
Lewis, F., 88
Lindl, J., 134, 157, 161
Linford, R., 101, 112, 163, 178
Lob, C., 131
Logan, G., 54, 67, 157, 165, 167, 219–220
Looney, P., 160

M
Manning, G., 69
Marburger, J., 154–156, 158, 159
Marmar, E., 134, 185
Marshall, C., 37
McBride, J., 34, 35, 41
McCormack, M., 34, 58, 59, 61–64, 72
McCrory, B., 112, 141, 165
McDaniel, P., 20, 21
McKnight, T.W., 134, 149–169
McLean, E., 27
Meade, D., 91, 121, 136, 139, 169, 222–223
Meserve, R., 143
Miley, G., 16
Miller, B.R., 32, 35, 36
Molvig, K., 99, 101
Mondale, W., 58
Moniz, E., 142
Monsler, M., 131
Montgomery, B., 129, 134
Mori, S., 55, 71
Moses, E., 165, 174
Moses, K.G., 35, 36
Motojima, O., 136, 176

N
Najmabadi, F., 96, 119, 152, 157
Navratil, G., 127, 152
Nixon, R., 28, 45
Nuckolls, J.H., 96, 223–224

O
Obama, B., 171–192
Obenschain, S., 167
Ohkawa, T., 21, 22, 33, 41, 59
Oki, N., 128
Oktay, E., 35, 36
O'Leary, H., 109, 116, 121, 123, 131, 136
O'Leary, J., 117
Olson, C., 157, 162, 167
O'Neill, P., 151
Orbach, R.L., 154, 156, 157

P
Pack, M., 108
Paidassi, S., 179
Palmieri, T., 64, 65
Papay, L., 127, 181
Parker, R., 39, 101, 152
Pastore, J., 21
Peacock, N., 23, 169
Pease, B., 23, 228
Peña, F., 137, 142
Peng, Y.-K.M., 88, 143

Perle, R.N., 89
Peterson, B., 56
Pewitt, D., 63–65
Porkolab, M., 127, 184
Post, R.F. (Dick), 5, 12, 21, 225–226
Post, R.S., 72
Postma, H., 29, 34, 42, 43
Prager, S., 127, 129, 134, 157, 161, 175, 184

R
Ray, D.L., 34, 45
Reagan, R., 63, 74, 99, 116
Rebut, P.-H., 115, 116
Repici, D., 64
Ribe, F., 34, 46, 72, 98, 101, 226–227
Richardson, B., 142, 143, 149
Rickover, A.H.G., 11–17, 19, 49, 51, 65, 205–209
Roberts, M. (Mike), 29, 34, 43, 44, 54, 88
Robinette, W., 131
Robinson, D., 23
Roddis, L., 72
Roderick, H., 20
Roe, R., 100
Rohrabacher, D., 129
Rose, D.J., 19, 20
Rosenbluth, M.N., 20, 21, 23, 34, 44, 62, 95, 107, 112, 114, 120, 132, 134, 225, 226
Rosner, B., 188
Rostoker, N., 21
Ruark, A., 20, 21
Rudakov, L., 39
Rutherford, E., 3
Rutherford, P., 33, 53, 116, 127

S
Sakharov, A., 5
Sauthoff, N., 129, 130, 134
Schlesinger, J., 28, 31, 52, 53, 56, 57
Schneider, T., 117, 119
Scott, B., 34, 35
Seaborg, G.T., 21, 25, 28
Seamans, B., 50
Seigel, K., 25
Sensenbrenner, J., 142
Sessler, A.M., 127, 181
Sethian, J., 146, 156, 157
Shaw, M., 49
Sheffield, J., 101, 112, 114, 127, 134, 145, 152, 157, 181, 228–229
Siemon, R., 127
Simonen, T., 96
Sleeper, A., 35, 41, 53
Smullin, L., 39
Sokolov, Y.A., 95
Sokolow, R., 127
Sooy, W., 27
Spencer, D., 118
Spitzer, L., 5, 12, 225
Stacey. B., 54, 55, 72
Stamper, J., 27
Steiner, D., 54
Stever, H.G., 107
Stuntz, L., 104
Synakowski, E., 175, 177, 178

T
Tamm, I., 5
Taschek, D., 34, 42
Taylor, J., 117
Teller, E., 5, 20, 110, 225, 226
Thomassen, K., 46
Thompson, G., 4
Thompson, J.R., 132, 133
Thonneman, P., 3
Todd, A., 131
Tomabechi, K., 95, 229–230
Tonks, L., 11
Toschi, R., 95
Toth, S., 131
Townes, C., 114
Trivelpiece, A.W., 31–35, 38, 41, 42, 58, 59, 64, 65, 67, 70, 81, 89, 94–96, 98, 120, 230–232
Tsongas, P., 62, 63, 66
Tuck, J., 4, 5, 226, 227

V
Van Atta, C., 34, 42, 48
Velikhov, E., 38, 88, 116, 128
Vest, C., 127

W
Ware, A., 4
Watkins, J.D., 99–101, 104–107, 109–111, 113, 116, 117, 120, 121
Watkins, R.A., 35, 36
Weinberg, A., 47
Wharton, C., 21
Whitman, C.T., 123
Wilcock, P., 23

Williams, J.M., 33, 35, 36, 46, 51, 54
Willis, E., 57
Willis, J., 35, 36, 96, 166, 167
Wirth, T., 98
Withers, G.K., 93
Wolfe, S., 77
Woodson, H., 70, 71
Wu, L., 127
Wurden, G., 5, 178

Y
Yamanaka, C., 69
Yergin, D., 123
York, H., 225
Young-sam, K., 128

Z
Zwilsky, K.M., 36

Subject Index

A

Advanced Reactor Innovations Evaluation Study (ARIES), 96, 119, 120
AEC, 20–25, 27–29, 31–35, 37, 39–41, 43–46, 49, 52, 53, 79, 121, 214, 225, 232
AEC policy and action paper, 21, 45
Alcator, 39, 57, 77, 85, 88, 91, 93, 97, 183–185, 187, 189
Alternate concepts, 32, 70, 112, 130, 139, 164, 177, 178
Amortizing fund, 106
An Accelerated Fusion Power Development Plan, 111
ARIES. *See* Advanced Reactor Innovations Evaluation Study (ARIES)
Astron, 21, 32, 33
Austin panel, 156
Availability, 25, 68, 73, 106, 118, 138, 196, 203, 228, 229, 233

B

Baseball II, 48
Blanket, 8, 73, 158, 194, 200, 201, 228, 229
BPAC. *See* Burning Plasma Assessment Committee (BPAC)
BPX. *See* Burning plasma experiment (BPX)
British thermal unit (BTU), 205, 206
Broader approach, 222
BTU. *See* British thermal unit (BTU)
Budget
  cuts, 53, 80, 81, 110–111, 131, 135, 139, 153, 166, 177, 235
  history, 85, 234
Bumpy Torus, 33, 34, 45, 47–48, 58, 67–69, 71, 75, 216
Burning Plasma Assessment Committee (BPAC), 157, 158, 160–163
Burning plasma experiment (BPX), 111, 112, 114, 128, 152, 153, 158, 160, 161, 222
Burning plasma panel, 153
Burning plasma physics, 152–154
Bush–Gorbachev summit meeting, 108

C

CDA. *See* Conceptual design activity (CDA)
Center for Fusion Engineering (CFE), 62–64
CIT. *See* Compact ignition tokamak (CIT)
Classification, 25, 26, 35, 37, 92, 93, 109, 110, 183
Climate change, 99, 143, 206, 208–209, 238
Coal, 56, 104, 193, 196, 205–208, 229
Collisionless shock waves, 24, 27
Columbus, 11
Compact ignition tokamak (CIT), 94–101, 104, 111, 114, 222
Complexity, 14, 111, 118, 178, 190, 201, 202
Conceptual design activity (CDA), 95, 116
Contract with America, 84, 126, 135
Controlled thermonuclear research (CTR), 20–25, 28, 29, 31, 42–44, 52, 53, 226, 232
Coulomb force, 2, 6, 7
Criteria for Practical Fusion Power Systems, 118
CTR. *See* Controlled thermonuclear research (CTR)

D

DCTR. *See* Division of controlled thermonuclear research (DCTR)
DCX-II, 20
Demonstration power plant, 31, 32, 51, 54, 85, 109, 115, 116, 131, 132, 141, 146–147, 157, 168, 179–180, 220

S.O. Dean, *Search for the Ultimate Energy Source: A History of the U.S. Fusion Energy Program*, Green Energy and Technology, DOI 10.1007/978-1-4614-6037-4,

Department of Energy (DOE), 32, 52–54, 56, 57, 61–67, 69–81, 87–89, 91–104, 106, 107, 109–121, 123, 125–135, 139, 142, 144–147, 149–164, 166–168, 172–177, 180–188, 190, 192, 195, 212, 214, 215, 217, 220, 223, 227, 229, 232–235
Desalination, 6, 196–197, 220, 237
Deuterium, 3–5, 7, 9, 24, 29, 34, 43, 81, 91, 98, 99, 102, 103, 115, 118, 119, 121, 126, 151, 162, 164, 174, 191, 201, 203, 221, 224, 228, 229, 235–237
Deuterons, 6, 43
Direct drive, 98, 103, 120, 156, 179, 182, 221, 222
Distinguished Career Awards, 167, 176
Divertor, 200, 201, 213
Division of controlled thermonuclear research (DCTR), 28, 29, 31, 33–37, 41, 42, 44, 46, 49, 50
DOE. *See* Department of Energy (DOE)

E

EDA. *See* Engineering design activities (EDA)
Efficiency, 68, 81, 97, 118, 137, 139, 150, 172, 209, 219, 224, 233
Electricity, 2, 12, 25, 106, 107, 127, 138, 143, 151, 193, 194, 196, 197, 200, 202–204, 206–209, 219–221, 230, 231, 237, 238
Electric power, 2, 6, 35, 67, 70, 74, 90, 116, 117, 121, 138, 161, 166, 181, 193–195, 199, 217–219, 230, 232
Electric Power Research Institute (EPRI), 35, 59, 67, 70, 74, 90, 112, 117–121, 181, 193, 194, 219
Electric utilities, 28, 46, 90, 118, 149, 193, 215
Electric vehicles, 106, 193, 208
Electrolysis, 194
EMC2, 16
Energy Policy Act of 2005, 165–166, 187
Energy Research Advisory Board (ERAB), 62, 71–73, 79, 104
Energy Research and Development Administration (ERDA), 45, 46, 49–52, 79, 219
Energy Technology Development Trust Fund, 106
Engineering, 15, 19, 32, 61, 88, 105, 130, 149, 174, 199, 211, 235
Engineering design activities (EDA), 112, 116, 133, 142
Engineering test facility (ETF), 54, 61, 62, 66, 79, 109, 151
EPRI. *See* Electric Power Research Institute (EPRI)
ERAB. *See* Energy Research Advisory Board (ERAB)
ERDA. *See* Energy Research and Development Administration (ERDA)
ETF. *See* Engineering test facility (ETF)
Experimental power reactor, 31, 54, 72, 147

F

FEAC. *See* Fusion Energy Advisory Committee (FEAC)
Feedback stabilization, 46
FESAC. *See* Fusion Energy Sciences Advisory Committee (FESAC)
FICUS. *See* Fusion Industry Council US (FICUS)
Field reversed concept (FRC), 16, 17, 19, 111, 227
FIRE. *See* Fusion Ignition Research Experiment (FIRE)
Fire from the Sun, 108
First wall, 199, 200
Fission, 1–3, 5, 6, 15, 19, 20, 40, 41, 49, 53, 106, 137, 150, 172, 193–197, 200, 201, 203, 204, 206–208, 214, 218–220, 229, 237, 238
Forum for Next Step Major Experiments, 141
Fossil energy, 6, 144
Foster panel, 53, 54
FPA. *See* Fusion power associates (FPA)
FPAC. *See* Fusion Policy Advisory Committee (FPAC)
FPCC. *See* Fusion Power Coordinating Committee (FPCC)
FRC. *See* Field reversed concept (FRC)
Fusion Energy Advisory Committee (FEAC), 112–114, 119, 130–134, 139
Fusion Energy Sciences Advisory Committee (FESAC), 132–134, 139, 145, 152, 153, 156, 157, 159–161, 163, 164, 168, 173, 175–178, 184–186, 188, 190, 192, 194, 213, 217, 229
Fusion–fission hybrid, 181, 194–195
Fusion Ignition Research Experiment (FIRE), 156, 160, 223

Fusion Industry Council US (FICUS), 130, 134
Fusion Policy Advisory Committee (FPAC), 107–110, 120, 121, 131, 132, 147, 214
Fusion policy statement, 73, 107
Fusion power associates (FPA), 54, 58–59, 63, 64, 66, 67, 69–71, 73–76, 78, 81, 85, 89, 90, 92, 101, 106, 107, 110–113, 117, 128, 130, 131, 134, 138, 142–143, 146, 149, 150, 167, 172, 175, 176, 179
Fusion Power Coordinating Committee (FPCC), 31–33, 44, 46, 68, 69, 213
Fusion systems, 61, 89, 151, 203, 204, 223
Fusion torch, 25, 196

G

General Atomics (GA), 20–23, 33, 34, 40–43, 47, 54, 55, 57, 59, 69, 74, 79, 84, 96, 97, 101, 102, 110, 112, 113, 130, 131, 134, 143, 150, 152, 154, 165, 181, 213
Goldston Scaling, 77

H

Halite–Centurion, 94, 110
HAPL. *See* High-average power laser (HAPL)
Heavy ion fusion (HIF), 97, 111, 165–167, 172, 220
HEDLP. *See* High-Energy-Density Laboratory Physics (HEDLP)
HIF. *See* Heavy ion fusion (HIF)
High-average power laser (HAPL), 146, 155–156, 167, 169, 173, 235
High-Energy-Density Laboratory Physics (HEDLP), 161, 168, 185
Hohlraum, 39, 98, 103, 162, 165, 174, 223, 224
Hydro, 74, 110, 206, 207, 238
Hydrogen, 1–4, 6, 8, 9, 42–44, 48, 56, 81, 84, 94, 151, 155, 166, 174, 194, 201, 220, 237
Hydrogen bomb, 4, 15, 25, 77, 237

I

ICC. *See* Innovative confinement concepts (ICC)
ICFAC. *See* Inertial Confinement Fusion Advisory Committee (ICFAC)
IEC. *See* Inertial electrostatic confinement (IEC)
IFE. *See* Inertial fusion energy (IFE)
Ignition, 7–9, 33, 40, 55–58, 73, 77, 79, 80, 88, 93, 99–101, 103, 120, 128, 137, 142, 144, 147, 153, 162, 165, 167, 168, 173, 174, 182, 183, 188, 189, 191, 219, 221, 222, 224, 231
Indirect drive, 39, 98, 103, 182, 221
Industrial participation, 70, 79, 83, 105, 215
Industry-Government Seminar on Fusion Energy Development, 70
Inertial confinement, 15, 16, 25, 37, 39, 58, 62, 69, 77–78, 84, 85, 87, 89, 91–94, 102–104, 107, 109, 110, 121, 126, 135, 139, 145, 146, 155, 165, 167, 168, 172–174, 178, 180, 182, 183, 189–191, 195, 201, 221, 224, 227, 229, 231
Inertial Confinement Fusion Advisory Committee (ICFAC), 120, 172
Inertial electrostatic confinement (IEC), 16, 17
Inertial fusion energy (IFE), 103, 109, 111, 112, 130, 141, 144, 145, 157, 159–163, 167–169, 171–174, 178, 180–183, 192, 195, 204, 211–213, 221–224, 228, 235
Inertial fusion energy workshop, 167–169
Innovative confinement concepts (ICC), 32, 153, 178
Instabilities, 11–13, 23, 44, 45, 103, 224–226
International collaboration, 84, 105, 121, 126, 165, 169, 189, 190, 211
International Thermonuclear Experimental Reactor (ITER), 14, 33, 77, 96–99, 107, 108, 112, 114–117, 121–123, 127, 128, 130, 132, 133, 136–138, 142, 149, 154–166, 172–177, 180, 183–192, 201, 202, 204, 213, 214, 216–219, 222, 223, 226, 227, 230, 231, 235–237
INTOR, 55, 72, 88, 95
Ion beams, 15, 144, 165, 182
ITER. *See* International Thermonuclear Experimental Reactor (ITER)
ITER Council, 116, 128, 174, 176, 191–192
ITER Industry Council, 77, 116, 130, 216

J

JASONS, 126, 173
JCAE. *See* Joint (House–Senate) Committee on Atomic Energy (JCAE)
JET. *See* Joint European Torus (JET)

Joint (House–Senate) Committee on Atomic Energy (JCAE), 21, 22, 42–43
Joint European Torus (JET), 33, 73, 84, 102, 115, 116, 121, 135, 136, 140, 152, 199, 201, 202, 223, 235
JT-60, 33, 73, 235

K
Kink instability, 11, 12
KMS, 25, 26, 59, 74, 77, 78, 110
Knotec panel, 131, 132
Krypton fluoride, 98, 172, 182
KSTAR, 136, 222

L
Laboratory Microfusion Facility (LMF), 97, 102–104, 120, 126
Large helical device (LHD), 36, 162, 176
Laser, 6, 15, 16, 24–27, 29, 69, 71, 74, 77, 78, 85, 93, 97, 98, 102, 103, 107, 114, 120, 126, 135, 146, 155, 156, 162, 165, 168, 172, 174, 179, 182, 189, 191, 195, 197, 199, 216, 221–224, 227, 231, 235–237
Laser fusion, 25–29, 41, 119, 156, 223
Lawson criterion, 7–8, 199
Lawson product, 77
Leadership awards, 63, 66, 167, 172, 175
Levitated dipole, 178, 185
LHD. *See* Large helical device (LHD)
Linear pinch, 11–13
Lithium, 5, 151, 200, 201, 228, 229
LLNL, 71, 74, 75, 94–96, 112, 114, 116, 125–127, 134, 141, 165, 174, 175
LMF. *See* Laboratory Microfusion Facility (LMF)

M
MAGLIF. *See* Magnetized liner inertial fusion (MAGLIF)
Magnetic bottles, 3, 11–15, 235
Magnetic Fusion Advisory Committee (MFAC), 70, 71, 73–74, 79, 80, 94, 97–101, 112, 213, 227
Magnetic Fusion Energy Engineering Act of 1980, 31, 61–63, 65, 72, 89–90, 116, 215
Magnetic Fusion Program Plan, 49–52
Magnetic mirror, 12, 13, 17, 20, 21, 33, 40, 45, 48–49, 57, 65, 68, 71, 75, 89, 95, 177, 213–216, 226, 227
Magnetized Liner Inertial Fusion (MAGLIF), 179
Magnetized target fusion (MTF), 16, 17, 178
Magneto-inertial fusion (MIF), 16, 178–179
Maintenance, 10, 12, 16, 68, 178, 200, 202–204, 218
Marketplace, 118, 123, 138, 143, 203
Massachusetts Institute of Technology (MIT), 19–21, 23, 34, 39, 53, 57, 59, 66, 70–73, 77, 80, 88, 91, 93, 96, 97, 99, 101, 107, 111, 112, 120, 127, 129, 130, 134, 136, 137, 150, 152, 153, 178, 181, 183–185, 187
Materials, 1, 6, 8, 10, 16, 25, 32–34, 36, 53, 68, 96, 97, 99, 109, 118, 126, 130, 137–139, 144, 151, 158, 163, 166, 182, 188, 191, 194–197, 199–201, 208, 211, 213, 218, 222, 223, 227–229
Megagauss magnetic fields, 27
MFAC. *See* Magnetic Fusion Advisory Committee (MFAC)
MFTE-B, 54, 65, 68–71, 74, 75, 80–82, 95, 214, 215, 227
MFTF. *See* Mirror Fusion Test Facility (MFTF)
MIF. *See* Magneto-inertial fusion (MIF)
Mirror Fusion Test Facility (MFTF), 54, 57, 68, 73
Mission, 66, 82, 112, 114, 120, 131–135, 146, 155, 173, 180, 190, 191, 212, 222, 230
MIT. *See* Massachusetts Institute of Technology (MIT)
MSNBC, 149
MTF. *See* Magnetized target fusion (MTF)

N
National center, 212
National Energy Policy Development Group (NEPD), 150–151
National energy strategy, 104–106, 111–112
National Ignition Facility (NIF), 15, 120–121, 125, 126, 130, 135, 139, 146, 156, 160, 165, 167–169, 173–176, 182, 189–192, 219, 221–224, 231, 235–237
National Resources Defense Council (NRDC), 126
Natural gas, 193, 194, 205–208, 219, 220, 229
NEPD. *See* National Energy Policy Development Group (NEPD)
Neutral beam, 9, 40, 43, 48, 49, 52, 54, 84, 121, 213

New York University, 20
NIF. *See* National Ignition Facility (NIF)
Nova laser, 93, 103, 223
NRDC. *See* National Resources Defense Council (NRDC)
Nuclear weapons, 4, 37, 45, 94, 110, 126, 135, 182, 201, 212, 223, 230, 231, 238

O
Office of Management and Budget (OMB), 21, 28, 29, 32, 50, 58, 63–66, 70, 74, 82, 89, 91, 97, 111, 114–116, 146, 171, 177, 233, 234
OHTE, 41
Oil, 6, 34, 74, 106, 123, 129, 150, 172, 205–208, 229
OMB. *See* Office of Management and Budget (OMB)
OMEGA, 135, 162, 165, 172, 179, 189, 191, 221
ORNL, 88, 112, 127, 134, 143, 152

P
Parallel Machine Scenario, 114
Pathways to Fusion Power, 138–139
PCAST. *See* President's Council on Science and Technology (PCAST)
Perhapsatron, 4, 5, 11, 227
PFC. *See* Plasma-facing components (PFC)
Physics Test Reactor, 30
Pilot plant, 111, 119
Pinch, 4, 11–15, 34, 41, 45–47, 71, 97, 111, 135, 177, 227
Pinch effect, 11
Plan, 29, 31, 32, 36, 40, 43, 45, 49–53, 55, 57, 61–85, 92, 98, 100, 101, 103, 104, 111, 114, 116, 117, 122, 123, 125, 127, 128, 133, 139, 141, 146, 147, 151, 155–157, 159–163, 167, 168, 172, 178, 184–186, 189, 190, 195, 215, 221, 223, 230, 234
Planning, 29–31, 33, 48–51, 63, 67, 68, 80, 84, 88, 89, 99, 102–104, 108, 112, 114, 118, 119, 131, 144, 145, 180–182, 185, 212, 236
Plasma, 4, 11, 19, 27, 61, 88, 105, 127, 150, 171, 194, 199, 211, 236
Plasma-facing components (PFC), 199, 200
PLT. *See* Princeton Large Torus (PLT)
Polar direct drive, 179
Politics, 233–235
PPPL. *See* Princeton Plasma Physics Laboratory (PPPL)
President's Council on Science and Technology (PCAST), 123, 127–130, 133, 137, 146, 147, 172
Princeton Large Torus (PLT), 29, 30, 33, 39, 40, 42, 43, 55, 57, 77
Princeton Plasma Physics Laboratory (PPPL), 21, 34, 77, 91, 95, 96, 99, 101, 107, 112, 114, 115, 121, 125, 127, 129–131, 134–136, 139, 143, 150, 157, 158, 165, 169, 172, 175, 179, 183, 213, 223
Priorities, 49, 53, 56, 70, 73, 79, 83, 88, 93, 97, 113, 114, 118, 133, 139, 145, 160, 161, 164, 185, 188, 189, 192, 200, 211–214, 233, 234
Progress, 10, 19, 25, 27, 33, 39, 46, 61, 66, 69, 71–73, 77, 78, 82, 84, 87, 92–94, 97, 100–102, 107, 108, 129, 133, 139, 144, 151–153, 156, 168, 169, 172, 182, 191, 211, 216, 221, 222, 224–228, 230, 231, 233, 235–237
Project Sherwood, 12, 13, 25, 55
Promise, 21, 23, 46, 47, 53, 61, 62, 88, 104, 107, 117, 121, 133, 139, 144, 151, 158, 159, 178, 183, 187, 195, 202, 211, 214, 217, 229, 234, 237–238
Pulsed power, 15, 39, 69, 78, 162, 175, 179, 182, 236

Q
Q enhancement, 53, 54

R
Radioactive waste, 6, 10, 107, 119, 151, 195, 200, 218, 238
Radioactivity, 6, 10, 34, 218
Reagan–Gorbachev summit meeting, 89–90, 95
Remote handling, 81, 135, 202, 204
Renewables, 6, 104, 106, 137, 150, 158, 183, 205, 208, 214, 219, 220, 237, 238
Reprocessing, 195, 208
Roadmap, 179–181, 203, 212

S
Sausage instability, 11
Scientific feasibility, 7, 25, 29, 102, 199
Scylla, 45, 46, 227
Scyllac, 45, 46, 57, 227

SEAB. *See* Secretary of Energy Advisory Board (SEAB)
SEAB fusion task force, 143
Secretary of Energy Advisory Board (SEAB), 114, 116, 143–145, 147, 173
Security, 24, 25, 35, 38, 39, 41, 73, 83, 88, 89, 106, 110, 158, 174, 182, 183, 188
Senior Committee on Environmental, Safety, and Economic Aspects of Magnetic Fusion Energy (ESECOM), 171
Solar, 1, 6, 50, 64, 104, 106, 193, 194, 203, 205, 206, 208, 216, 219, 220, 228, 238
Special Awards, 167
Spheromak, 16, 111
Spinoffs, 6, 66, 98, 197, 215
SSC. *See* Superconducting supercollider (SSC)
Staged theta pinch, 46
Standing Committee, 22–25, 31, 43, 44
Status and Objectives of Tokamak Systems for Fusion Research, 33, 43
Stellarator, 12–15, 21–24, 28, 39, 48, 71, 110, 136, 162, 177, 223, 225, 227, 228
Strategy, 49, 55, 64, 71, 73–74, 104–106, 109, 111–113, 115, 117, 130, 131, 137, 152, 157, 168, 178, 189, 212, 213, 220
Strong management, 143, 145, 159
Superconducting supercollider (SSC), 114, 215
Supply and demand, 209
Systems, 1, 24, 31, 61, 89, 105, 131, 151, 174, 193, 199, 211
Systems engineering, 37, 105, 199, 202, 203, 212

T

T-3, 23, 28, 39, 169, 227
TAERF. *See* Texas Atomic Energy Research Foundation (TAERF)
Tandem mirror, 54, 55, 67, 226
Technical planning activity (TPA), 88
Texas Atomic Energy Research Foundation (TAERF), 20, 22, 28
TFTR. *See* Tokamak Fusion Test Reactor (TFTR)
Thermal barriers, 54, 67–69
Theta pinch, 29, 33, 45, 46, 53, 227
Three Mile Island, 66, 67
Tokamak, 14, 22–25, 28–29, 61, 88, 111, 127, 152, 173, 200, 213, 235
Tokamak Fusion Test Reactor (TFTR), 30, 33, 37, 43–45, 51, 52, 57, 71, 73, 79–82, 84, 91, 93, 95, 97, 99, 112, 115, 118, 121, 128–130, 135–137, 139–140, 147, 152, 175, 199, 213, 214, 223, 233, 235
Tokamak Physics Experiment (TPX), 113, 116, 117, 122, 123, 125, 127, 128, 130, 135–136, 222
Tokamak Physics Test Reactor, 30
Tore Supra, 88, 135
Toroidal pinch, 11–14, 22, 48
TPA. *See* Technical planning activity (TPA)
TPX. *See* Tokamak Physics Experiment (TPX)
Tritium, 3–5, 7, 9, 24, 29, 34, 42–44, 53, 80, 81, 95, 98, 99, 102, 103, 115, 118, 119, 121, 126, 135, 140, 164, 174, 191, 200–203, 213, 221, 224, 228–230, 235
Tritons, 6–8
Tri-Valley Cares, 126

U

University of Maryland, 24, 28, 31, 32, 45, 51, 58, 111, 129
Utility requirements for fusion, 67, 68
UWMAK, 54

V

Vacuum, 20, 39, 48, 91, 135, 192, 199

W

Washington Post, 56
Waste, 6, 25, 68, 92, 118, 119, 194–196, 200, 206, 208, 218, 224, 237
Weapons, 6, 15, 29, 35, 68, 77, 78, 87, 89, 91–94, 98, 102, 103, 109, 110, 126, 130, 134, 139, 146, 155, 173, 188, 223, 233–235
Wet wood burner, 43, 44
Wind, 6, 38, 43, 56, 193, 194, 203, 205, 206, 208, 219, 220, 238
Workforce development, 217

X

2X-II, 48, 49

Z

Zeta, 11, 229
Z-pinch, 15, 45, 46, 58, 155, 162, 168, 226, 227

CPSIA information can be obtained
at www.ICGtesting.com
Printed in the USA
BVOW07*1332070416
443348BV00014B/12/P

9 781461 460367